anatomy of
MUSCLE
BUILDING

anatomy of
MUSCLE
BUILDING

Craig Ramsay

A FIREFLY BOOK

Published by Firefly Books Ltd. 2011

First printing

Publisher Cataloging-in-Publication Data (U.S.)

Ramsay, Craig
    Anatomy of muscle building : a trainer's guide to increasing muscle mass / Craig Ramsay ; Jonathan Conklin.
[160] p. : col. ill., col. photos. ;   cm.
Summary: A step-by-step guide to exercises that help build the muscles of the chest, back, legs, arms, and abs, with anatomical illustrations that point out the relevant muscles. Also contains tips on performing each exercise correctly and lists of all the muscles used in each exercise.
ISBN-13: 978-1-55407-825-7 (bound)
ISBN-10: 1-55407-825-3
ISBN-13: 978-1-55407-816-5 (pbk.)
ISBN-10: 1-55407-816-4
1. Bodybuilding. 2. Muscles—Anatomy. I. Conklin, Jonathan. II. Title.
613.713 dc22   GV546.5R367 2011

Library and Archives Canada Cataloguing in Publication

Ramsay, Craig
    Anatomy of muscle building : a trainer's guide to increasing muscle mass / Craig Ramsay.
ISBN-13: 978-1-55407-825-7 (bound)
ISBN-10: 1-55407-825-3
ISBN-13: 978-1-55407-816-5 (pbk.)
ISBN-10: 1-55407-816-4
    1. Bodybuilding. 2. Muscles--Anatomy. I. Title.
GV546.5.R35 2011   613.7'13   C2010-906015-6

Published in the United States by
Firefly Books (U.S.) Inc.
P.O. Box 1338, Ellicott Station
Buffalo, New York 14205

Published in Canada by
Firefly Books Ltd.
66 Leek Crescent
Richmond Hill, Ontario L4B 1H1

Printed in Canada

CONTENTS

INTRODUCTION

Whether you are a competitive bodybuilder or just someone who wants to look sculpted and fit, your focus is on toning and building key muscles.

In the following pages, you will find a selection of muscle-building exercises divided into major target areas: chest and abdominals, back, shoulders, arms, and legs. Step-by-step instructions accompany the photos of the exercises, along with anatomical illustrations that show you the major active and stabilizing muscles that are being worked. Active muscles are those that contract to move a body part or structure, and stabilizing muscles either co-contract or activate to stabilize the active muscle.

An understanding of which muscles are working during a given exercise will help you better craft a workout regimen to fit your goals.

CHOOSING THE RIGHT GYM

So you've decided to embark on a muscle-building regimen, and now the question arises: Where should I work out? These days, there are so many choices of facilities—from high-tech health clubs that feature the latest in exercise machines to bare-bones gyms that offer little more than free weights. A gym is an extension of your goals—a hard-core body builder is looking for a different environment than the financial executive who needs a place to unwind after spending 12 hours at the office. Here is a rough outline that will help you match a gym to your specific fitness goals.

CHOOSING THE RIGHT GYM

All gyms are not created equal—each one caters to a different clientele. Just what kind of clientele a gym is trying to attract determines what type of equipment it offers, what hours it operates, and how much it charges. You need to refine your goals so that you can determine whether a particular gym is the right one for you. When choosing a gym, make sure you start with a trial membership so you can be sure it is the right fit for you, without any obligation or commitment. Check out exactly what it offers before signing any membership contracts.

Want to build muscle mass?

If your goal is to build muscle mass, look for a gym with the right quality and quantity of the weight equipment you need. A muscle-building gym will have plenty of cable, Smith, and bench

press machines, along with weight plates and free weights. Check out the quality of the machines, too. For example, are the machine attachments made of steel, or are they covered in rubber? The most efficient muscle-building weight machines will have steel attachments that facilitate the use of wrist straps—you cannot properly tighten wrist straps on rubber-covered equipment.

Many of these muscle-building gyms also feature juice or snack bars that offer high-energy foods like high-protein smoothies and power bars loaded with complex carbs.

Want to obtain a lean, tight, sculpted body?

Look for a gym with a wide variety of cardio and circuit machines such as, treadmills, stationary bikes, ellipticals,

GYM ETIQUETTE

- Don't rush around a gym. With all that heavy equipment, there is so much that can really injure you. Move slowly but with purpose.

- If you have questions, ask! Leave your ego at the door. Gym equipment can be very dangerous. Protect yourself by understanding how a machine works. Ask a gym representative to explain how to correctly use each piece of gym equipment.

- Take advantage of personal trainers. They can inform you about the ins and outs of the gym and which programs and pieces of equipment you should focus on in order to get closer to your goals.

- Avoid blocking the mirrors or standing in front of a dumbbell or other weight rack.

- Share equipment and weights. Don't sit on a machine in between sets. "Work in" with other gym members.

- Remove weight plates from a machine or bar when you finish with them, and be sure to put the weights back where you found them.

- Wipe off equipment when you are finished, so you don't "sweat out"' your neighbor. Be mindful of others.

- Use the lockers. Most gyms have a policy that gym bags cannot be carried around the gym. Gym bags and other comparable material can be potential tripping hazards.

- Time manage your workout. Most gyms impose a time limit on using cardio equipment so that all members can take advantage of it. Use common sense—if the gym is busy, limit your time on a specific cardio machine to 20 minutes. When the gym is less crowded, you can go for a bit longer.

INTRODUCTION

stepping machines, stair climbers, and arm bikes, along with equipment like kettle bells, foam rollers, bosu balance trainers, medicine balls, and fitness balls. Bodybuilding and weightlifting equipment should be present, too, but limited to such basics as barbells, dumbbells, and cable machines. More important, the gym should offer a wide variety of classes that will motivate you to keep active. These class offerings may include, spinning, dance, cardio burn, abdominal training, and intensive boot camps, just to name a few.

If you are a woman and a newcomer to the gym, or perhaps are intimidated by the social aspects of the gym environment, consider a women's-only gym. These types of gyms should have an educated staff to help make you feel comfortable and confident with your workout program.

BEFORE YOU JOIN

Other things to think about when choosing the right gym:

- Cleanliness
- Location
- Pricing
- Staff attitude and qualifications
- Type of clientele
- Hours of operation

Before you sign on the dotted line, ask the gym representative a few important questions: Can your membership be frozen, refunded, or transferred? Does the gym have injury insurance or passport programs (meaning you can use the franchise locations while traveling)? Also, make sure to bargain with your membership representative. There is plenty of competition out there, and each company wants your business—you have this fact on your side when negotiating a membership deal.

LIFTING AIDS: YES OR NO?

Power lifters tend to use aided equipment, such as lifting belts and lifting wrist straps to reduce the chance of injury when lifting weights that are heavier than normal.

Use of a weight belt and wrist straps should be handled with care, and with the understanding

that they are made to aid extra-heavy lifting. Wrist straps stabilize your grip, allowing you to hold onto a heavier weight without it slipping out of your hands. A weightlifting belt generally does two things: It reduces stress on the lower back when you are upright, and it prevents back hyperextension during overhead lifts.

A belt also has the benefit of keeping you aware of your back position, but there are many fitness experts who question the need for belts for all but the heaviest lifters. In fact, the use of a weight belt can actually weaken your core, because a belt inhibits the muscles that work to stabilize the abdomen. Still, individuals who have suffered an injury and are in need of added support usually wear wrist and knee wraps. Be sure to consult a doctor for any such recommendations.

BUILDING A TRAINING PROGRAM

There are many factors that come into play when building a training program—it all depends on your current fitness level, and your ultimate goals. When working with my own clients, I follow a basic formula that takes into account their fitness levels.

• Beginners: Start with four sets of 12 to 20 repetitions per exercise. The first set should be a lower-weight warm-up. For the second, third, and fourth sets, use appropriate beginner's weight.

• Advanced: Use a four-set program with the first one being an easy warm-up of 16 reps per exercise. For the second, third, and fourth sets, use heavy weights with six to eight reps per exercise.

When deciding how heavy to lift, pick a weight that allows you to complete your target reps only, bringing you near to muscle failure. In general, lower rep ranges (6 to 10 reps) with heavier weights will add mass; higher rep ranges (10 to 20 reps) with lighter weights are best for toning muscles without adding too much bulk.

WEIGHT-LIFTING TIPS

• Perfect your form—proper form when lifting is your best defense against injury.
• Keep your spine in neutral position, and whenever lifting or setting down weights, use your leg muscles, not your back.
• Don't go it alone. When lifting heavy weights, find a spotter.
• Dress properly. Make sure you clothing is breathable and allows freedom of movement. Shoes with good traction will help keep you grounded and stable.

FULL-BODY ANATOMY

scalenus*

sternocleidomastoideus

pectoralis major

pectoralis minor*

deltoideus anterior

serratus anterior

coracobrachialis*

biceps brachii

rectus abdominis

obliquus internus*

obliquus externus

pronator teres

palmaris longus

flexor digitorum*

flexor carpi ulnaris

extensor carpi radialis

transversus abdominis*

flexor carpi pollicis longus

flexor carpi radialis

tensor fasciae latae

sartorius

iliopsoas*

vastus intermedius*

iliacus*

rectus femoris

pectineus*

vastus lateralis

adductor longus

vastus medialis

gracilis*

tibialis anterior

gastrocnemius

peroneus

soleus

extensor hallucis

flexor digitorum

adductor hallucis

extensor digitorum

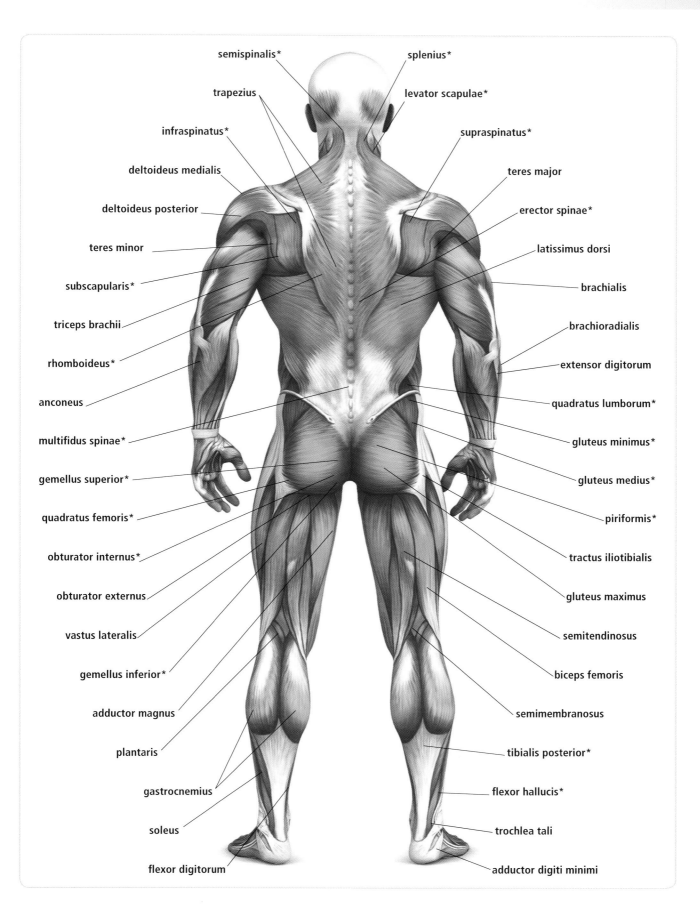

semispinalis*

splenius*

trapezius

levator scapulae*

infraspinatus*

suprasipinatus*

deltoideus medialis

teres major

deltoideus posterior

erector spinae*

teres minor

latissimus dorsi

subscapularis*

brachialis

triceps brachii

brachioradialis

rhomboideus*

extensor digitorum

anconeus

quadratus lumborum*

multifidus spinae*

gluteus minimus*

gemellus superior*

gluteus medius*

quadratus femoris*

piriformis*

obturator internus*

tractus iliotibialis

obturator externus

gluteus maximus

vastus lateralis

semitendinosus

gemellus inferior*

biceps femoris

adductor magnus

semimembranosus

plantaris

tibialis posterior*

gastrocnemius

flexor hallucis*

soleus

trochlea tali

flexor digitorum

adductor digiti minimi

WARM-UPS

Although you may be tempted to walk into the gym and head straight from the locker room to the weight racks, warming up is essential to any exercise program, including muscle-building regimens. Warm-ups will increase the benefits of exercising and help decrease the potential of sustaining injury. The basic kinds of warm-ups fall into two categories: cardiovascular exercises and stretches. Cardiovascular exercises stimulate blood and oxygen flow through your body. Stretches gradually and smoothly lengthen the muscles, maximizing their flexibility.

The verdict is out on when—or even if—muscle builders should stretch. Before? After? Both? Not at all? Whichever routine you choose, to get the most from a stretch, only stretch to a point where you feel a slight tension—stretches should not be painful. Never bounce into a deeper stretch; controlled movement is the key to a safe, beneficial stretch.

CARDIO WORKOUTS

The cardiovascular system, which includes the heart, lungs, and blood vessels, is the circulatory system that distributes blood throughout our bodies. Cardiovascular exercise aims to increase the heart rate, making oxygen and nutrient-rich blood available to working muscles.

When you perform cardio exercises, the energy burned by the working muscles elevates the blood temperature, which causes your heart to start beating faster. Your blood vessels, which dilate from the higher blood temperature, allow more blood and oxygen to reach your muscles, making them more elastic and less susceptible to injury. Joint activity also becomes easier, increasing the efficiency of any exercise regimen.

TARGET
• Cardiovascular system

JUMP ROPE

You do not need to perform a lot of cardio before a muscle-building workout. Short periods of from 5 to 15 minutes sufficiently prepare you for some heavy lifting. Just keep in mind that a good cardio session will allow you to maintain a heart rate no higher than 120 beats per minute.

Most gyms are equipped with a variety of cardio machines, such as treadmills, rowing machines, stationary

BOSU JUMPS

bicycles, stair machines, and elliptical trainers. If you are unfamiliar with a machine's functions, ask a trainer or gym representative how to use it. You can also opt for a low-tech cardio workout, such as bouncing or running in place on a trampoline, jumping on a bosu ball, or jumping rope.

TRAMPOLINE BOUNCES

CHEST & FRONT DELTOID STRETCH

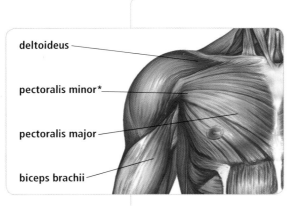

deltoideus

pectoralis minor*

pectoralis major

biceps brachii

ANNOTATION KEY

* indicates deep muscles

1 Stand straight with your arms behind your back and your hands clasped together.

2 Pinch your shoulder blades together as you reach and lift your arms away from your body, making sure to keep your elbows straight.

3 Hold for 15 seconds before returning your arms back to the starting position. Repeat.

TARGET
- Chest
- Shoulders

MUSCLES USED
- pectoralis major
- pectoralis minor
- deltoideus anterior
- biceps brachii

TRAPEZIUS STRETCH

① Standing with your feet parallel and shoulder width apart, gently grasp the side of your head with your right hand.

trapezius

ANNOTATION KEY

* indicates deep muscles

MUSCLES USED

- scalenus
- sternocleidomastoideus
- trapezius

TARGET
- Upper back

② Tilt your head toward your raised elbow until you feel the stretch in the side of your neck.

③ Turn your head toward your right shoulder, as you continue to feel the stretch.

④ Hold for 15 seconds, and repeat. Switch sides, and repeat sequence on left side.

sternocleidomastoideus

scalenus*

SHOULDER STRETCH

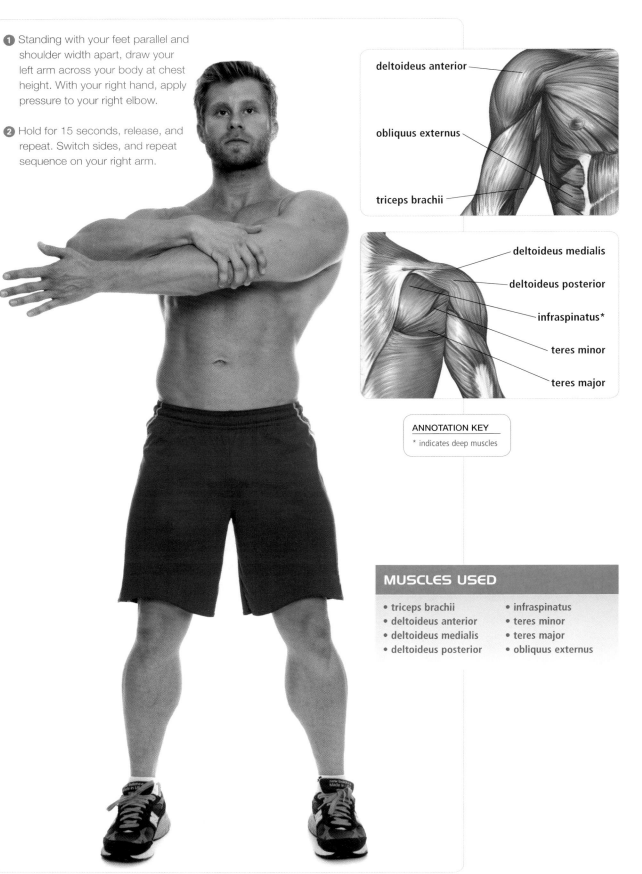

1 Standing with your feet parallel and shoulder width apart, draw your left arm across your body at chest height. With your right hand, apply pressure to your right elbow.

2 Hold for 15 seconds, release, and repeat. Switch sides, and repeat sequence on your right arm.

TARGET
• Deltoids

deltoideus anterior

obliquus externus

triceps brachii

deltoideus medialis

deltoideus posterior

infraspinatus*

teres minor

teres major

ANNOTATION KEY
* indicates deep muscles

MUSCLES USED

• triceps brachii
• deltoideus anterior
• deltoideus medialis
• deltoideus posterior
• infraspinatus
• teres minor
• teres major
• obliquus externus

TRICEPS STRETCH

① Standing with your feet parallel and shoulder width apart, raise your left arm and bend it behind your head.

② Keeping your shoulders relaxed, gently pull on your left elbow with your right hand.

③ Continue to pull on your elbow until you feel the stretch in your lower shoulder. Hold for 15 seconds and repeat. Switch sides, and repeat sequence with your right arm bent.

MUSCLES USED

- **triceps brachii**
- **deltoideus posterior**
- **infraspinatus**
- **teres major**
- **teres minor**

deltoideus posterior

infraspinatus

teres minor

teres major

triceps brachii

ANNOTATION KEY
* indicates deep muscles

TARGET
- Triceps

FOREARM STRETCH

1 Standing with your feet parallel and shoulder width apart, bend your left arm, keeping your upper arm close to your side, and your palm facing inward.

2 With your right hand, gently push your left hand in toward your torso.

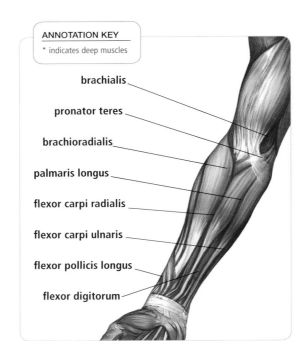

ANNOTATION KEY
* indicates deep muscles

brachialis

pronator teres

brachioradialis

palmaris longus

flexor carpi radialis

flexor carpi ulnaris

flexor pollicis longus

flexor digitorum

TARGET
• Lower arm muscles

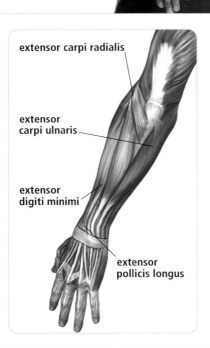

extensor carpi radialis

extensor carpi ulnaris

extensor digiti minimi

extensor pollicis longus

3 Turn your left hand so that your palm is now facing out, and again apply light pressure with your right hand.

4 Switch sides, and repeat sequence on your right hand.

MUSCLES USED

- brachialis
- pronator teres
- brachioradialis
- palmaris longus
- flexor carpi radialis
- flexor carpi ulnaris
- flexor pollicis longus
- flexor digitorum
- extensor carpi radialis
- extensor carpi ulnaris
- extensor digiti minimi
- extensor pollicis longus

TOE TOUCH

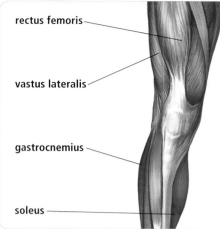

rectus femoris

vastus lateralis

gastrocnemius

soleus

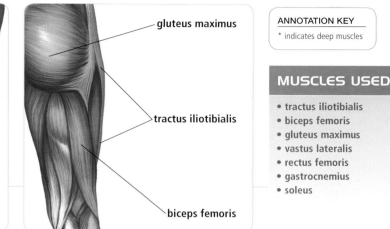

gluteus maximus

tractus iliotibialis

biceps femoris

ANNOTATION KEY

* indicates deep muscles

MUSCLES USED

- **tractus iliotibialis**
- **biceps femoris**
- **gluteus maximus**
- **vastus lateralis**
- **rectus femoris**
- **gastrocnemius**
- **soleus**

① Stand with your feet parallel and shoulder width apart and your arms at your sides.

② Bend from the hips while keeping both knees straight, reaching your hands toward the floor.

③ Hold for 15 seconds, and repeat.

TARGETS
- Iliotibial band
- Hamstrings

ⓐ

ⓑ

TRAINER'S TIPS

- The tractus iliotibialis, or iliotibial band (ITB), is a thick band of connective tissue that crosses the hip joint and extends down to insert on the kneecap, tibia, and biceps femoris tendon. The ITB stabilizes the knee and abducts the hip.

SPINAL STRETCH

❶ Lie on your back with your legs straight and arms outstretched at shoulder height.

❷ Bend your right leg, and keeping both shoulders on the floor, slowly bring your leg across your body until you feel the stretch in the area between your lower back and hips.

❸ Stretch only as far as your shoulders will allow without one of them rising from the floor.

MUSCLES USED

- quadratus lumborum
- erector spinae
- multifidus spinae
- vastus lateralis
- tractus iliotibialis
- tensor fasciae latae
- latissimus dorsi

TARGET
- Back muscles

❹ Hold for 15 seconds, and repeat. Switch legs and repeat the entire sequence with your left leg bent.

ANNOTATION KEY

* indicates deep muscles

tensor fasciae latae

vastus lateralis

erector spinae*

latissimus dorsi

multifidus spinae*

quadratus lumborum*

tractus iliotibialis

LOWER- AND UPPER-BACK STRETCH

supraspinatus*

infraspinatus*

teres minor

subscapularis*

teres major

latissimus dorsi

erector spinae*

quadratus lumborum*

multifidus spinae*

semitendinosus

biceps femoris

semimembranosus

1 Sit on the floor or an exercise mat with your legs straight out, your ankles bent at a 90-degree angle so that your toes point toward the ceiling.

2 Loosely clasp your hands, and rest your forearms on your knees, bending your torso forward from the hips.

3 Without bouncing, continue to lean forward, concentrating on stretching your entire spine.

4 Hold your lowest point for about 15 seconds, and repeat.

ⓐ

TARGETS
- Back muscles
- Hamstrings

ANNOTATION KEY
* indicates deep muscles

MUSCLES USED

• supraspinatus	• erector spinae
• infraspinatus	• quadratus lumborum
• teres minor	• multifidus spinae
• subscapularis	• semitendinosus
• teres major	• biceps femoris
• latissimus dorsi	• semimembranosus

ⓑ

HAMSTRING STRETCH

1. Standing with your feet parallel and shoulder width apart, extend your right leg forward.

2. Bend your left knee as you tilt your hips forward, and rest both hands on your right knee. Your weight should be on your left, bent knee.

3. Hold for between 10 and 30 seconds, and repeat. Switch legs and repeat the entire sequence with your right leg.

gltuteus maximus

semitendinosus

biceps femoris

semimembranosus

ANNOTATION KEY
* indicates deep muscles

TARGET
• Hamstrings

MUSCLES USED

• biceps femoris
• semitendinosus
• semimembranosus
• gluteus maximus

HAMSTRING-ADDUCTOR STRETCH

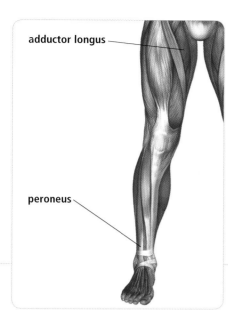

adductor longus

peroneus

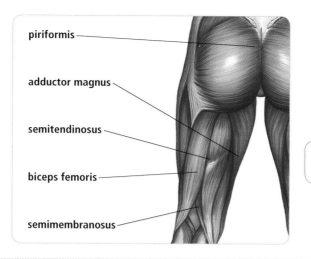

piriformis

adductor magnus

semitendinosus

biceps femoris

semimembranosus

ANNOTATION KEY
* indicates deep muscles

① Stand with your feet planted well beyond shoulder width apart, so that you are in a straddle position. Bend your knees.

② Place both hands on your left knee, keeping your spine in neutral position and your shoulders slightly forward.

③ Keeping your torso in the same position and your hips behind your heels, shift your weight to the left, bending your left knee while extending your right leg. Hold for 10 seconds and repeat on other side.

MUSCLES USED

- adductor longus
- adductor magnus
- peroneus
- biceps femoris
- semitendinosus
- semimembranosus
- piriformis

TARGETS
- Hamstrings
- Inner thighs

QUAD STRETCH

tensor fasciae latae

rectus femoris

vastus lateralis

vastus medialis

TARGET
• Quadriceps

1 Stand with your feet parallel and close together, with your arms at your sides.

2 With your right hand, reach behind as you bend your right knee. Grasp your right foot and gently pull your heel toward your buttocks with your hand until you feel a stretch in the front of your thigh. Keep both knees together and aligned.

3 Hold for 15 seconds, and repeat. Switch sides, and repeat on the left leg.

MUSCLES USED

• rectus femoris   • vastus medialis
• vastus lateralis   • tensor fasciae latae

LUNGE STRETCH

iliopsoas*

iliacus*

pectineus*

tensor fasciae latae

sartorius

adductor longus

rectus femoris

gracilis*

ANNOTATION KEY
* indicates deep muscles

MUSCLES USED

- iliopsoas
- iliacus
- rectus femoris
- sartorius
- tensor fasciae latae
- pectineus
- adductor longus
- gracilis

❶ Stand with your legs straight, feet close together and your arms at your sides.

a

TARGET
- Hip flexors

❷ Step forward with your right leg, placing both hands on your right knee as you bend it to lower your pelvis to the ground.

❸ Keeping your head up, shoulders level, and eyes looking forward, hold for 15 seconds. Repeat sequence three times on each leg.

b

GROIN STRETCH

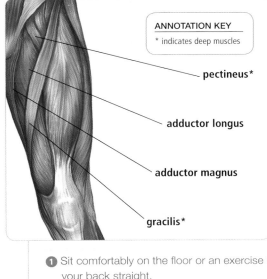

ANNOTATION KEY

* indicates deep muscles

pectineus*

adductor longus

adductor magnus

gracilis*

MUSCLES USED

- adductor longus
- adductor magnus
- pectineus
- gracilis

TARGET
- Groin muscles

1 Sit comfortably on the floor or an exercise mat with your back straight.

2 With your knees pointing outward, lightly grasp the tops of your feet as close to the toes as is comfortable, and bring the soles of your feet together.

3 Without bouncing, keep your back straight and gently press your knees toward the floor, feeling a stretch in your inner thighs.

4 Hold 15 to 30 seconds, and repeat.

SEATED GLUTE STRETCH

1 Sit comfortably on the floor or an exercise mat with your back straight.

2 Clasp both hands around your right knee, and bring your right foot over your left thigh, resting your ankle on your thigh.

3 At the same time, bend your left knee and draw your foot inward so that that the side of your left foot is resting on the floor, close to the underside of your raised right thigh.

ANNOTATION KEY

* indicates deep muscles

gluteus minimus*

gluteus medius*

piriformis*

gluteus maximus

a

4 Hold for 15 seconds, and switch sides. Repeat sequence with your left ankle crossed resting on your right knee.

TARGET
• Gluteal muscles

MUSCLES USED

• gluteus maximus
• gluteus medius
• gluteus minimus
• piriformis

b

SUPINE GLUTE STRETCH

1 Lie on your back with your legs elevated and knees bent in a tabletop position.

2 Bring your right ankle over your left knee, resting it on your left thigh. Place both hands around your left thigh.

MUSCLES USED
- piriformis
- gluteus maximus
- gluteus medius
- gluteus minimus

a

TARGET
- Gluteal muscles

3 Gently pull your thigh toward your chest until you feel the stretch in your buttocks. Hold for 15 seconds, and switch sides. Repeat sequence with your left ankle crossed resting on your right knee.

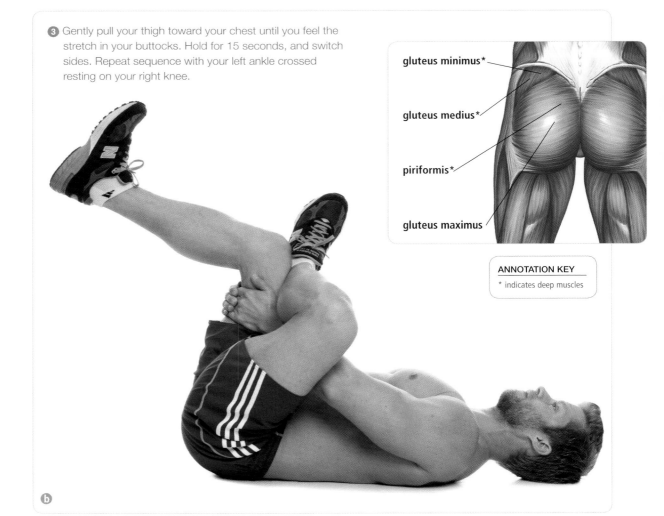

gluteus minimus*

gluteus medius*

piriformis*

gluteus maximus

ANNOTATION KEY

* indicates deep muscles

b

CALF STRETCH

1 Stand with your feet parallel and close together, with your arms at your sides. Place a dumbbell on the floor in front of you.

2 Step forward to place the toes of your left foot on the dumbbell bar.

3 Lower your heel to the floor until you can feel a stretch.

4 Hold for 20 to 30 seconds, and repeat. Switch sides, and repeat on the right leg.

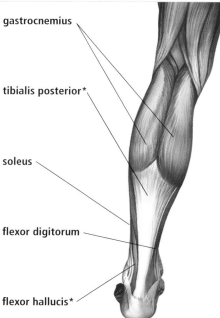

gastrocnemius

tibialis posterior*

soleus

flexor digitorum

flexor hallucis*

ANNOTATION KEY

* indicates deep muscles

MUSCLES USED

- gastrocnemius
- tibialis posterior
- soleus
- flexor digitorum
- flexor hallucis

TARGET
- Calf muscles

CHEST & ABDOMINALS

The large chest muscles, the pectoralis major and minor, draw the arms forward and in toward the center of the body. The "pecs" also work with the shoulders and arms to perform pushing movements. The serratus anterior originates on the surface of the upper ribs at the side of the chest. Sometimes called the "big swing muscle," it pulls the scapula forward and around the rib cage, a move that occurs when someone throws a punch.

The abdominal muscles, the rectus abdominis and transversus abdominis, are located on the lower midsection of the torso. The "abs" contract the body forward. The side abdominals, the obliquus externus, are located on each side of the rectus abdominis. Lying beneath the external oblique is the obliquus internus. The obliques flex the rib cage and the pelvic bones together, bend the torso sideways, and rotate the torso. Targeted working of these muscles—along with a proper diet—results in the highly defined look known as "six-pack" or "washboard" abs.

BALANCE BALL CRUNCH

1 Lie back on a balance ball with your shoulders and head hanging off the ball, keeping your knees and hips bent. Gently hyperextend your back to conform to the contour of the ball.

2 Place your hands on the sides of your head, with your elbows bent.

a

TARGETS
• Upper abdominals

LOOK FOR
• Your abdominal muscles to initiate the movement.
• Your pelvis to remain in neutral position during the crunching motion.
• Your neck to remain elongated and relaxed.

AVOID
• Rocking back and forth. Keep your back stable on the ball.
• Holding your breath—empty your lungs and take catch breaths when necessary.
• Lifting your shoulders to help raise your torso.

3 Flex your waist to raise your upper torso.

4 Return to the starting position, and repeat.

b

MUSCLES USED

- rectus abdominis
- obliquus externus
- obliquus internus
- transversus abdominis

ANNOTATION KEY

Black text indicates active muscles

Gray text indicates stabilizing muscles

* indicates deep muscles

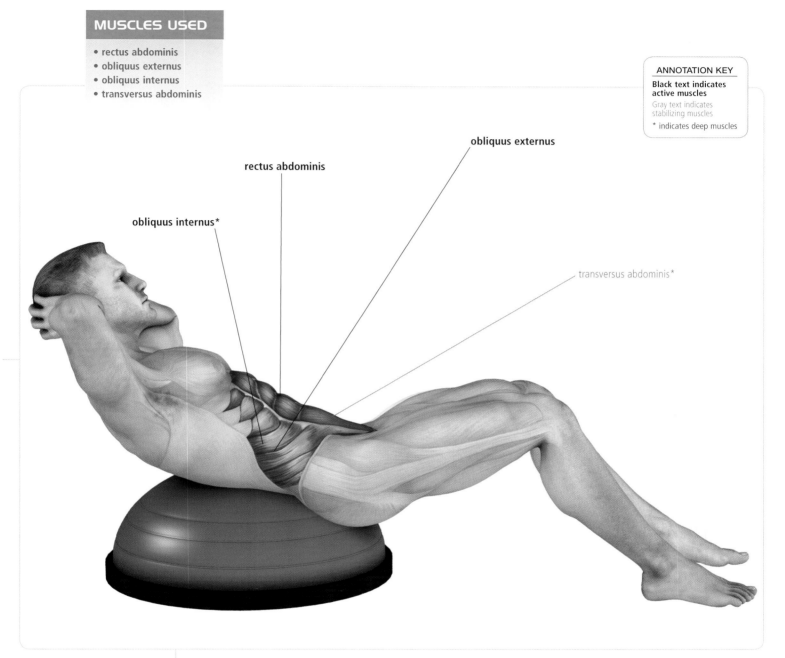

obliquus externus

rectus abdominis

obliquus internus*

transversus abdominis*

TRAINER'S TIPS

- Keep your feet planted firmly on the floor.
- If you feel neck strain, take a small towel and place it behind the occipital bone of your head. Grip the ends with each hand, and keeping your elbows in, lift your chin toward the ceiling.
- A bosu ball works best, but if none are available, use a stabilizing ball.

ROPE PULL-DOWN

① Kneel on the floor in front of a cable machine.

② Grasp the cable rope attachment with both hands.

a

③ Bend forward from the waist, pull downward on the cable, and place your wrists against your head.

④ Flex your hips so that the resistance on the cable pulley lifts your torso upward so that your spine hyperextends.

b

TRAINER'S TIPS
• Do not hold your breath. Let air out of your lungs, and take short catch breaths when necessary.
• Do not use arm strength to execute movement—focus on using your abdominal muscles.

⑤ Keeping your hips stationary, flex your waist so that your elbows move toward the middle of your thighs. Return to starting position, and repeat.

c

TARGETS
• Upper abdominals
• Obliques

LOOK FOR
• Movement to occur in your waist.

AVOID
• Shifting your hips once movement begins.

MUSCLES USED

• rectus abdominis
• obliquus internus
• obliquus externus
• serratus anterior
• iliopsoas
• tensor fasciae latae
• rectus femoris
• sartorius
• latissimus dorsi
• teres major
• deltoideus posterior
• triceps brachii
• rhomboideus
• trapezius
• pectoralis major
• pectoralis minor

MODIFICATION
Similar difficulty:
Follow steps 1 through 4, and then as you bend forward from the waist, twist to one side, moving your elbow toward the middle of the opposite thigh.

MODIFICATION
More difficult:
Follow previous instructions, but deepen the twist, aiming your elbow for the opposite knee.

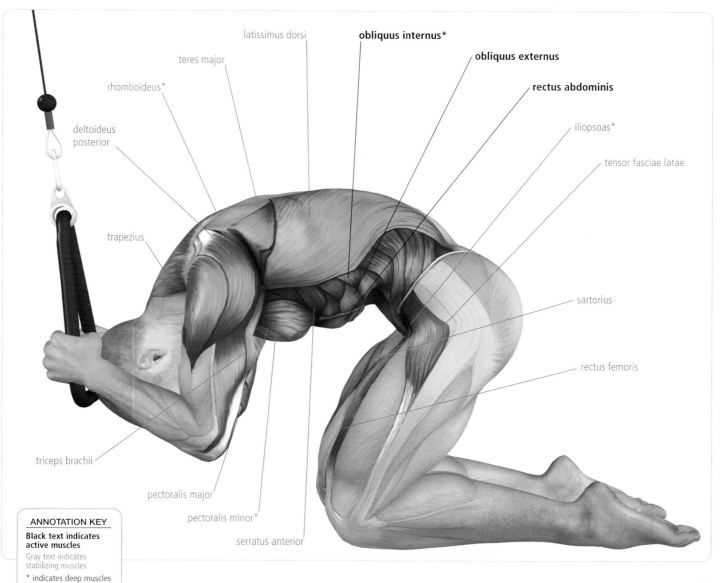

latissimus dorsi

obliquus internus*

teres major

obliquus externus

rhomboideus*

rectus abdominis

deltoideus posterior

iliopsoas*

tensor fasciae latae

trapezius

sartorius

rectus femoris

triceps brachii

pectoralis major

pectoralis minor*

serratus anterior

ANNOTATION KEY
Black text indicates active muscles
Gray text indicates stabilizing muscles
* indicates deep muscles

DOUBLE-CABLE LEAN-OVER

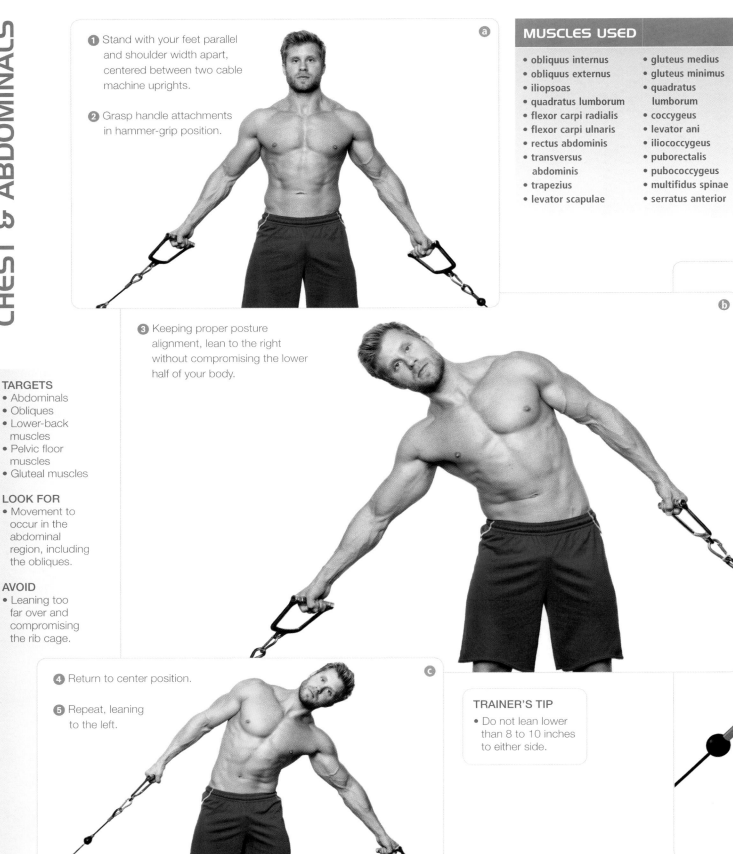

a

1. Stand with your feet parallel and shoulder width apart, centered between two cable machine uprights.

2. Grasp handle attachments in hammer-grip position.

MUSCLES USED

- obliquus internus
- obliquus externus
- iliopsoas
- quadratus lumborum
- flexor carpi radialis
- flexor carpi ulnaris
- rectus abdominis
- transversus abdominis
- trapezius
- levator scapulae
- gluteus medius
- gluteus minimus
- quadratus lumborum
- coccygeus
- levator ani
- iliococcygeus
- puborectalis
- pubococcygeus
- multifidus spinae
- serratus anterior

b

3. Keeping proper posture alignment, lean to the right without compromising the lower half of your body.

TARGETS
- Abdominals
- Obliques
- Lower-back muscles
- Pelvic floor muscles
- Gluteal muscles

LOOK FOR
- Movement to occur in the abdominal region, including the obliques.

AVOID
- Leaning too far over and compromising the rib cage.

c

4. Return to center position.

5. Repeat, leaning to the left.

TRAINER'S TIP
- Do not lean lower than 8 to 10 inches to either side.

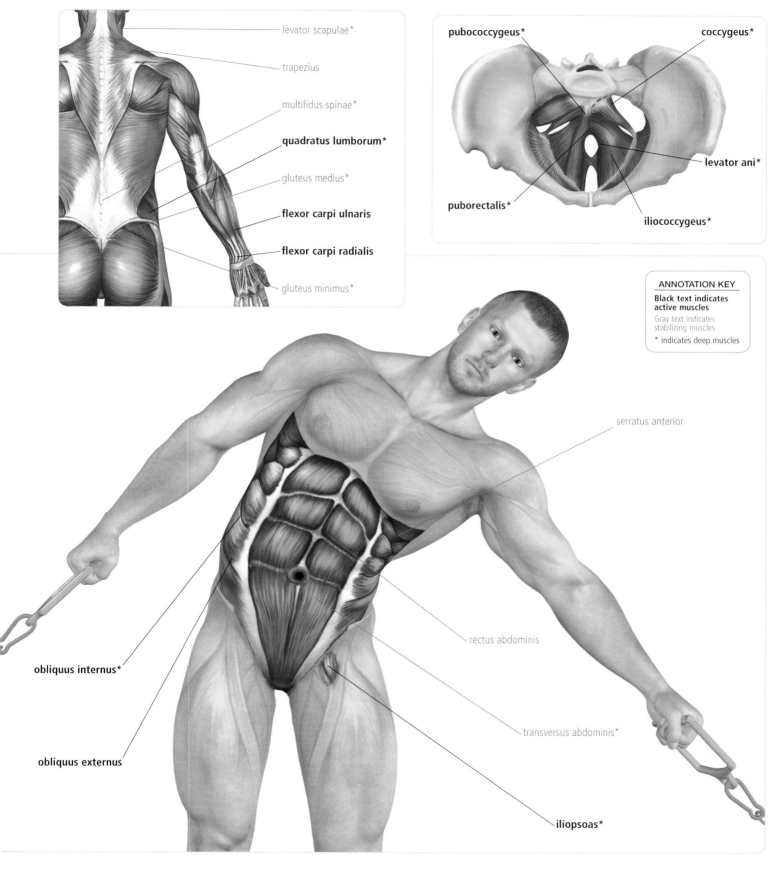

levator scapulae*

trapezius

multifidus spinae*

quadratus lumborum*

gluteus medius*

flexor carpi ulnaris

flexor carpi radialis

gluteus minimus*

pubococcygeus*

coccygeus*

levator ani*

puborectalis*

iliococcygeus*

ANNOTATION KEY

Black text indicates active muscles

Gray text indicates stabilizing muscles

* indicates deep muscles

serratus anterior

rectus abdominis

obliquus internus*

obliquus externus

transversus abdominis*

iliopsoas*

CABLE FLY

1 Stand between two high cable machine uprights. Grasp overhead handle grips in each hand, one at a time.

2 Center yourself between the cable uprights.

3 Take a full step back, bringing your hands toward your thighs.

a

b

MUSCLES USED

- pectoralis major
- pectoralis minor*
- rhomboideus*
- levator scapulae*
- deltoideus anterior
- latissimus dorsi
- biceps brachii
- brachialis
- triceps brachii
- flexor carpi radialis
- flexor carpi ulnaris
- rectus abdominis
- obliquus externus
- obliquus internus
- erector spinae
- serratus anterior

TARGET
- Upper chest

LOOK FOR
- Your hands to continue to face each other in the hammer-grip position.
- Your arms to be fully extended throughout the movement.

AVOID
- Extending your arms too far back—this will compromise your technique and could lead to a rotator cuff injury.

4 Step forward and start exercise with your hands facing each other, just below the chest. Place one leg in front of the other and slightly lunge forward, putting your weight on your front foot.

5 Extend your arms backward and out to the side until you feel a slight stretch in your chest.

c

d

TRAINER'S TIPS
- Start off with a light weight until you have mastered the movement and feel confident that you have the strength to execute it.
- Keep a slight bend in your elbow. This will ease the stress on your shoulder joint.

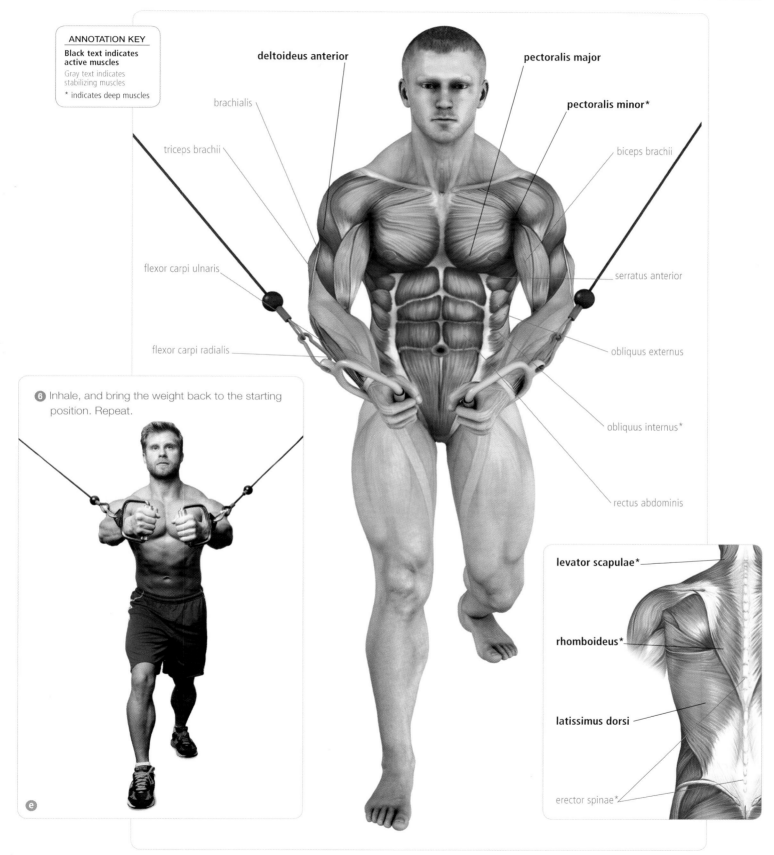

ANNOTATION KEY
Black text indicates active muscles
Gray text indicates stabilizing muscles
* indicates deep muscles

deltoideus anterior

brachialis

triceps brachii

flexor carpi ulnaris

flexor carpi radialis

pectoralis major

pectoralis minor*

biceps brachii

serratus anterior

obliquus externus

obliquus internus*

rectus abdominis

6 Inhale, and bring the weight back to the starting position. Repeat.

e

levator scapulae*

rhomboideus*

latissimus dorsi

erector spinae*

REVERSE-GRIP INCLINE BENCH PRESS

1 Using an incline bench, place dumbbells on your thighs, and then kick them up while leaning back on the bench.

2 Start with your elbows near your ribs, the dumbbells under your chest, and your palms facing behind you.

3 Lift the dumbbells straight up, making a half twist of the wrist so that your palms are facing forward.

4 Lower the dumbbells, making a half twist, returning you to your starting position.

5 Bring the dumbbells back to your thighs.

TARGET
• Upper chest

LOOK FOR
• Your elbows to stay close to your ribs. Grip the dumbbells to ensure that your wrists are in the proper position.
• Both feet to remain planted on the floor.

AVOID
• Lifting your glutes or back off the bench to aid the movement.
• Lifting your shoulders toward your ears.
• Hyperextending your arms at the top range of movement.

TRAINER'S TIP
• Relax your jaw, and breathe out as you lift the dumbbells up, and breathe in as you lower them.

MUSCLES USED

- pectoralis major
- deltoideus anterior
- triceps brachii
- biceps brachii

deltoideus anterior

biceps brachii

pectoralis major

triceps brachii

ANNOTATION KEY

Black text indicates active muscles

Gray text indicates stabilizing muscles

* indicates deep muscles

HAMMER-GRIP PRESS

1 Sit on an incline bench, grasping a pair of dumbbells on your thighs.

a

b

2 Kick up while leaning back on the bench, so that you can start the movement with your elbows near your ribs and the dumbbells to the side of your chest. Your palms should be facing each other in hammer-grip position.

TARGET
- Upper chest

LOOK FOR
- Isolation of the chest muscles.
- The dumbbells to face each other as you execute the movement.

AVOID
- Lifting your feet off the floor.
- Lifting your glutes and back while executing the exercise.
- Hyperextending your arms at the top range of the movement.

3 Raise the dumbbells toward the ceiling until they are directly above your shoulders. Hold this top position for a moment.

4 Slowly lower the dumbbells to your shoulders, and then bring the dumbbells back to the starting position. Repeat.

c

d

MUSCLES USED

- pectoralis major
- pectoralis minor
- deltoideus anterior
- triceps brachii

TRAINER'S TIPS
- Proper wrist position is essential—this one is like "punching" wrists.
- Relax your jaw and neck muscles.
- Exhale as you lift the dumbbells up, and inhale as you lower them down.
- Keep your elbows in while executing the exercise.

pectoralis minor*

triceps brachii

deltoideus anterior

pectoralis major

ANNOTATION KEY

Black text indicates active muscles

Gray text indicates stabilizing muscles

* indicates deep muscles

PLATE PUSHUP

1 Lie prone on floor with your hands slightly wider than shoulder width apart and your fingertips parallel to your collarbone. Place both feet on your toes, about 6 to 8 inches apart, and raise your body off the floor by extending your arms, pushing into the floor as you keep your body straight.

TRAINER'S TIPS
- Use your abdominals to support yourself.
- Focus on an object that is about 2 feet in front of you.
- Push yourself to muscle failure.

a

TARGETS
- Upper, middle, and lower chest

LOOK FOR
- Both upper and lower body to remain straight throughout the movement.
- Your pelvis to be slightly tucked during the movement.
- Your chest to be lifted.

AVOID
- Pointing your elbows to the side—they should be facing the back corners behind you.
- Hyperextending your arms at the top range of the movement.

2 Have a training partner place a 45-pound weight plate on the upper region of your back, in between the scapulas. If necessary, place a towel on the upper to middle back region—this will prevent the weight from irritating the skin and causing discomfort.

MUSCLES USED
- pectoralis major
- deltoideus anterior
- triceps brachii
- biceps brachii
- rectus abdominis
- obliquus internus
- obliquus externus
- rectus femoris
- vastus lateralis
- vastus intermedius
- vastus medialis
- pectoralis minor
- serratus anterior

b

3 Slowly lower your body by bending your arms, maintaining a single plane from shoulder to feet.

4 Push back up into the starting position, and repeat.

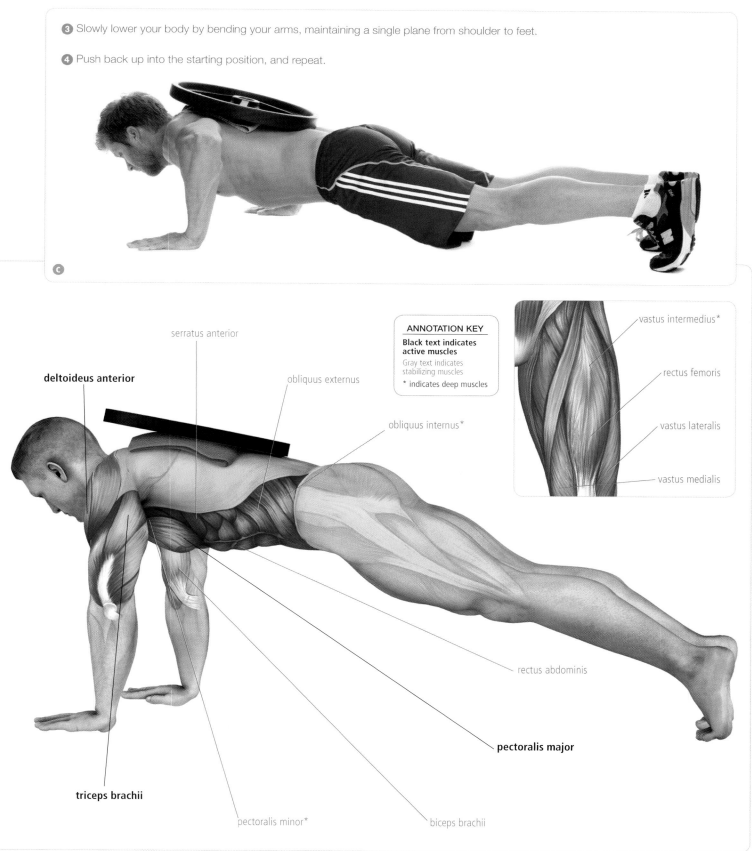

serratus anterior

deltoideus anterior

obliquus externus

obliquus internus*

ANNOTATION KEY

Black text indicates active muscles

Gray text indicates stabilizing muscles

* indicates deep muscles

vastus intermedius*

rectus femoris

vastus lateralis

vastus medialis

rectus abdominis

pectoralis major

triceps brachii

pectoralis minor*

biceps brachii

BENT-ARM DUMBBELL PULL-OVER

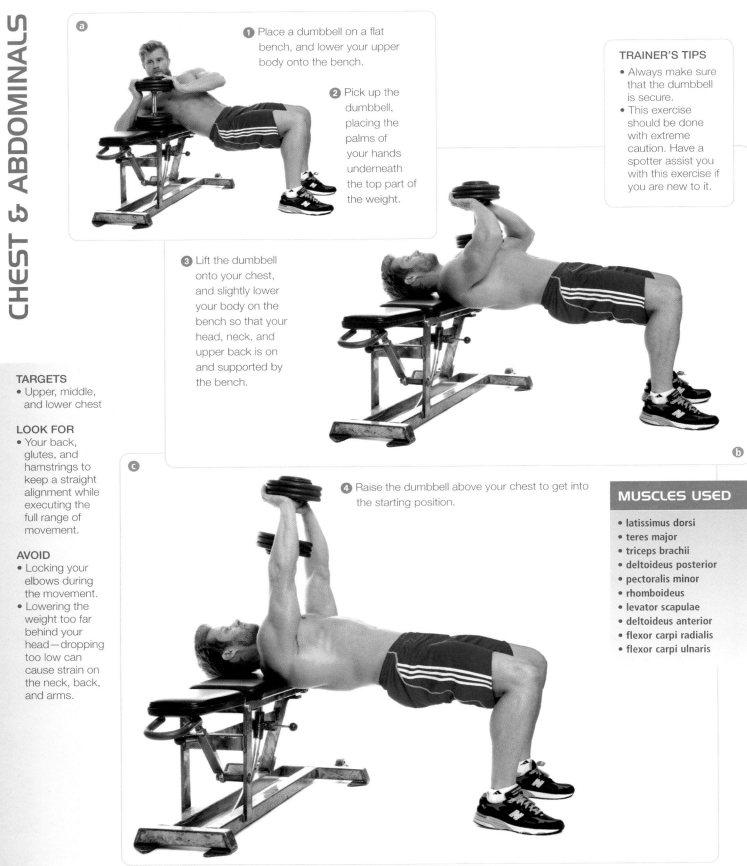

CHEST & ABDOMINALS

a

1. Place a dumbbell on a flat bench, and lower your upper body onto the bench.

2. Pick up the dumbbell, placing the palms of your hands underneath the top part of the weight.

3. Lift the dumbbell onto your chest, and slightly lower your body on the bench so that your head, neck, and upper back is on and supported by the bench.

b

c

4. Raise the dumbbell above your chest to get into the starting position.

TARGETS
• Upper, middle, and lower chest

LOOK FOR
• Your back, glutes, and hamstrings to keep a straight alignment while executing the full range of movement.

AVOID
• Locking your elbows during the movement.
• Lowering the weight too far behind your head—dropping too low can cause strain on the neck, back, and arms.

TRAINER'S TIPS
• Always make sure that the dumbbell is secure.
• This exercise should be done with extreme caution. Have a spotter assist you with this exercise if you are new to it.

MUSCLES USED

• latissimus dorsi
• teres major
• triceps brachii
• deltoideus posterior
• pectoralis minor
• rhomboideus
• levator scapulae
• deltoideus anterior
• flexor carpi radialis
• flexor carpi ulnaris

5 With your elbows slightly bent, lower the dumbbell over and behind you. Do not lower the dumbbell lower than your head.

6 Carefully return to the starting position.

levator scapulae*

rhomboideus*

flexor carpi radialis

flexor carpi ulnaris

deltoideus anterior

pectoralis minor*

triceps brachii

deltoideus posterior

teres major

latissimus dorsi

ANNOTATION KEY

Black text indicates active muscles

Gray text indicates stabilizing muscles

* indicates deep muscles

DUMBBELL FLY

ⓐ

① Grasping a dumbbell in each hand, sit on an incline bench with your shoulders higher than your hips. Place dumbbells on thighs to start.

ⓑ

② Lie back on an incline bench, kicking up dumbbells with your elbows in as you lift them to shoulder height.

③ Lift the dumbbells above your chest with your palms facing each other in hammer-grip position. Keep your elbows bent and your shoulder blades contacting the bench.

ⓒ

④ Keeping your spine in a neutral position, place your feet flat on the floor. Raise the dumbbells above your chest until your elbows are only very slightly bent.

TARGET
• Middle chest

LOOK FOR
• Your chest and rib cage to rise as the dumbbells descend.
• Your spine and shoulders to remain in the same position as you return to the starting position.
• Your elbows to be on a horizontal plane, even with the bench, when you reach the lowest position.

AVOID
• Moving your head or chin forward or off the bench.
• Elevating your shoulders.
• Bending your elbows excessively as the dumbbells descend, or flattening them as the dumbbells ascend.

MUSCLES USED

• pectoralis major
• deltoideus anterior
• biceps brachii
• coracobrachialis
• brachialis
• triceps brachii
• flexor carpi radialis
• flexor digitorum
• extensor digitorum
• brachioradialis
• extensor carpi radialis
• serratus anterior
• rectus abdominis
• deltoideus posterior
• subscapularis

TRAINER'S TIPS
• Kick into starting position.
• When done right, this move feels as if you are hugging a large tree.
• Keep your grip strong and your upper arms, both biceps and triceps, contracted.

ⓓ

5 Keeping your elbows bent, push your hands apart, and inhale until your hands drop to just below the height of your chest. Return to the starting position by squeezing your chest and bringing the dumbbells back along the same path as the descent, exhaling as you do so.

deltoideus posterior

subscapularis*

ANNOTATION KEY
Black text indicates active muscles
Gray text indicates stabilizing muscles
* indicates deep muscles

deltoideus anterior

brachialis

flexor carpi radialis

extensor carpi radialis

brachioradialis

flexor digitorum*

biceps brachii

triceps brachii

extensor digitorum

coracobrachialis*

pectoralis major

serratus anterior

rectus abdominis

SMITH MACHINE FLAT BENCH PRESS

1 Lie supine on a Smith machine bench with your upper chest positioned under the bar.

2 Grasp the barbell with an overhand grip. Disengage the bar by rotating it backward or forward.

a

MUSCLES USED

- pectoralis major
- deltoideus anterior
- triceps brachii
- coracobrachialis
- biceps brachii

b

3 Lower the barbell to your chest, and then press upward until your arms are fully extended.

4 Return to starting position, and repeat.

TARGET
- Middle chest

LOOK FOR
- Consistent motion.
- Your rib cage to remain open and rise during the descent phase.
- Your shoulders to remain retracted and away from your ears during the ascent phase.

AVOID
- Bouncing the weight off your chest.
- Positioning your hands too far apart—this will compromise your range of motion.
- Hyperextending your arms at the top range of movement—this will take the weight off the contracting muscles.

deltoideus anterior

pectoralis major

biceps brachii

triceps brachii

coracobrachialis*

ANNOTATION KEY

Black text indicates active muscles

Gray text indicates stabilizing muscles

* indicates deep muscles

TRAINER'S TIP

• To keep yourself stable while lifting, make sure that your head, shoulders, and hips remain in contact with the bench throughout the movement.

CABLE MACHINE DECLINE RAISE

1 Stand between two high cable machine uprights, your feet shoulder width apart and your pelvis tucked in. Pick up handle grips one at a time.

2 Center yourself between the cable uprights.

3 Start with your arms at the sides of your body, your elbows pointing to the wall behind you and your palms facing forward. You should feel a slight stretch in the chest.

a

TRAINER'S TIP
• Keep your shoulders down and back and your chest elevated.

b

4 Begin the movement by drawing your hands up and away from your body.

TARGET
• Lower chest

LOOK FOR
• Your knees to bend slightly.
• Your jaw to remain relaxed.

AVOID
• Collapsing your chest during the exercise.
• Rotating your palms inward when bringing the weight back to the starting position.

5 Bring the cables up into a pyramid position, the handles coming together just above your naval and continuing to slightly above your chest.

6 Lower the weight back to the starting position. Repeat.

c

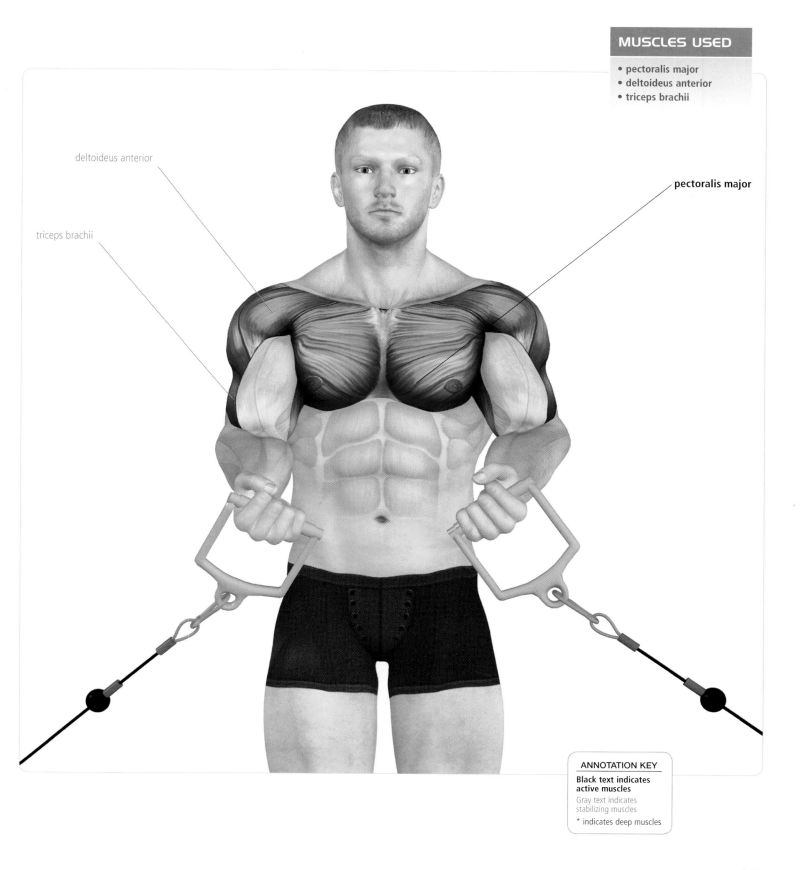

MUSCLES USED

- pectoralis major
- deltoideus anterior
- triceps brachii

deltoideus anterior

triceps brachii

pectoralis major

ANNOTATION KEY

Black text indicates active muscles

Gray text indicates stabilizing muscles

* indicates deep muscles

CABLE CROSSOVER FLY

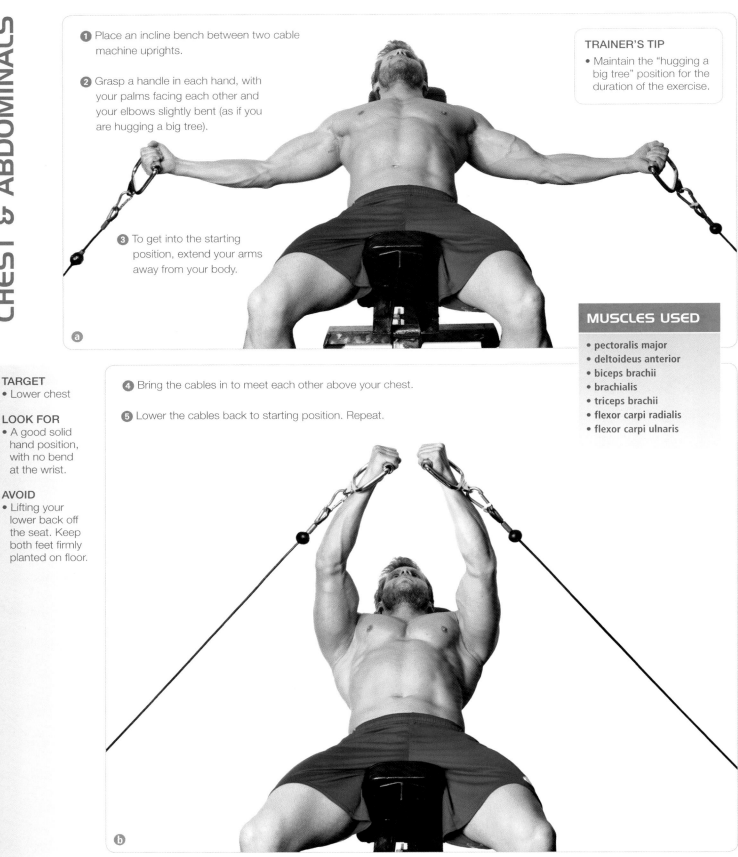

1 Place an incline bench between two cable machine uprights.

2 Grasp a handle in each hand, with your palms facing each other and your elbows slightly bent (as if you are hugging a big tree).

3 To get into the starting position, extend your arms away from your body.

TRAINER'S TIP
• Maintain the "hugging a big tree" position for the duration of the exercise.

MUSCLES USED
• pectoralis major
• deltoideus anterior
• biceps brachii
• brachialis
• triceps brachii
• flexor carpi radialis
• flexor carpi ulnaris

TARGET
• Lower chest

LOOK FOR
• A good solid hand position, with no bend at the wrist.

AVOID
• Lifting your lower back off the seat. Keep both feet firmly planted on floor.

4 Bring the cables in to meet each other above your chest.

5 Lower the cables back to starting position. Repeat.

MODIFICATION

Similar difficulty: Follow steps 1 and 2, but start the exercise with your palms facing forward and a 90-degree bend in your arms. Press the cables up to meet each other above your chest. Lower the cables to the starting position, and repeat. This exercise is called a Seated Cable Chest Press.

ANNOTATION KEY

Black text indicates active muscles

Gray text indicates stabilizing muscles

* indicates deep muscles

biceps brachii

deltoideus anterior

pectoralis major

flexor carpi ulnaris

flexor carpi radialis

brachialis

triceps brachii

CABLE DECLINE FLY

1 Adjust two cables so they are set on highest setting. Grasp one handle, and then reach out for the other handle.

2 Center yourself between the cable uprights. Stand with your feet placed shoulder width apart, your knees slightly bent.

3 With your elbows slightly bent and your palms facing downward, pull your arms down at an angle to meet in front of your body.

4 Lower the weights to starting position, and repeat.

TARGET
• Lower chest

LOOK FOR
• Your shoulders to remain down and away from your ears, allowing your chest to elevate.

AVOID
• Leaning too far forward.
• Swaying the lower back.

TRAINER'S TIP
• Exhale as you lower the weight, and inhale as you return it to the starting position.

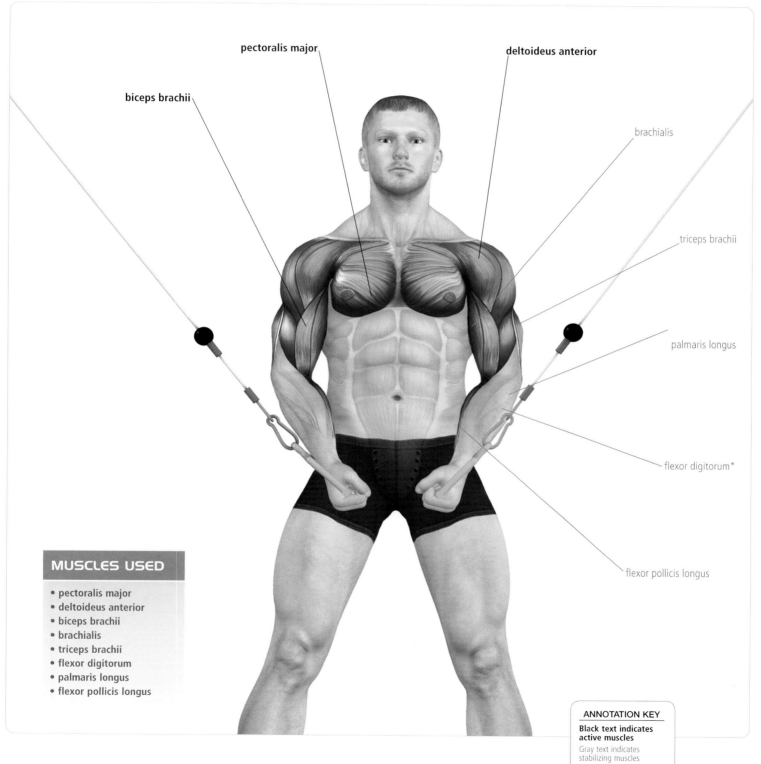

biceps brachii

pectoralis major

deltoideus anterior

brachialis

triceps brachii

palmaris longus

flexor digitorum*

flexor pollicis longus

MUSCLES USED

- pectoralis major
- deltoideus anterior
- biceps brachii
- brachialis
- triceps brachii
- flexor digitorum
- palmaris longus
- flexor pollicis longus

ANNOTATION KEY

Black text indicates active muscles

Gray text indicates stabilizing muscles

* indicates deep muscles

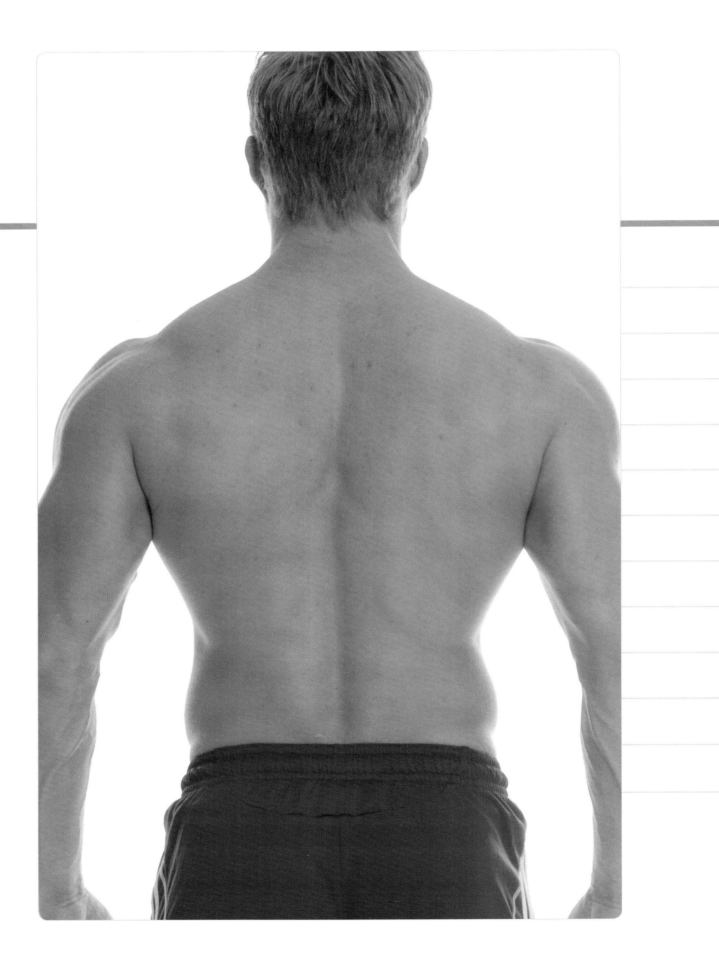

BACK

The back muscles are essential to daily life, moving the spine, hips, head, arms, and pelvis. This group includes the latissimus dorsi muscles, or "lats" as they are often called. The lats are the large, fan-shaped muscles located on the back of the torso that function to pull the arms down and back. Well developed lats give bodybuilders and weight lifters their characteristic V-shaped torsos. The trapezius, or "traps," is the flat, triangular muscle that covers the back of the neck, shoulders and thorax. The upper part of the trapezius elevates the shoulder and braces the shoulder girdle when a weight is carried.

Other major back muscles are the teres major, erector spinae, quadratus lumborum, and multifidus spinae.

INCLINE BENCH ROW

a

1 Holding a dumbbell in each hand, straddle an incline bench, facing toward the back.

2 Lean forward and carefully place the dumbbells on the bench.

b

3 Grasping the dumbbells with your palms facing each other in hammer-grip position, roll the dumbbells off the bench as you carefully lower your body until your chest rests against the bench.

4 Keeping your elbows close to the sides of your body, lift them toward the ceiling to draw the dumbbells upward.

TARGET
• Back

LOOK FOR
• Your chest to remain elevated throughout the exercise.
• Your feet to remain firmly planted on the floor.

AVOID
• Rushing this exercise.
• Using momentum to lift the dumbbells.
• Holding onto neck and jaw tension.
• Sliding down the bench during exercise.

5 Lower the dumbbells to starting position, and repeat.

c

TRAINER'S TIPS
• Use caution and care when executing this exercise.
• When getting into the starting position, place your pelvis on the bench first, then your stomach, and then your chest upon the dumbbells. Be sure to replace the dumbbells with your chest in a controlled manner.
• To protect your back and shoulders, carefully drop the dumbbells when you have finished the exercise.

trapezius

infraspinatus*

teres minor

rhomboideus*

teres major

latissimus dorsi

ANNOTATION KEY

Black text indicates active muscles

Gray text indicates stabilizing muscles

* indicates deep muscles

MODIFICATION

Similar difficulty: Follow instructions for the Incline Bench Row, but hold the dumbbells with your hands facing behind you in.

a

b

deltoideus posterior

brachialis

triceps brachii

pectoralis major

biceps brachii

brachioradialis

MUSCLES USED

- trapezius
- rhomboideus
- latissimus dorsi
- teres major
- deltoideus posterior
- infraspinatus
- teres minor
- brachialis
- brachioradialis
- pectoralis major
- biceps brachii
- triceps brachii

DUMBBELL SHRUG

 1 Holding a dumbbell in each hand, stand with your feet placed shoulder width apart and your knees slightly bent.

a

b

TARGET
• Upper back

LOOK FOR
• Your chest to lift during the upward movement.
• Your jaw to remain relaxed throughout the exercise.
• Your elbows to point behind you.

AVOID
• Bending your elbows—lift your shoulders up toward your ears.

2 Lift your shoulders straight up toward your ears.

3 Slowly and deliberately lower your shoulders to starting position, and repeat.

TRAINER'S TIPS
• Exhale when lifting your shoulders to your ears, and inhale while lowering the dumbbells.
• When finishing the exercise, use caution in releasing the dumbbells. Bend and use your legs to lower the dumbbells—do not use your back.

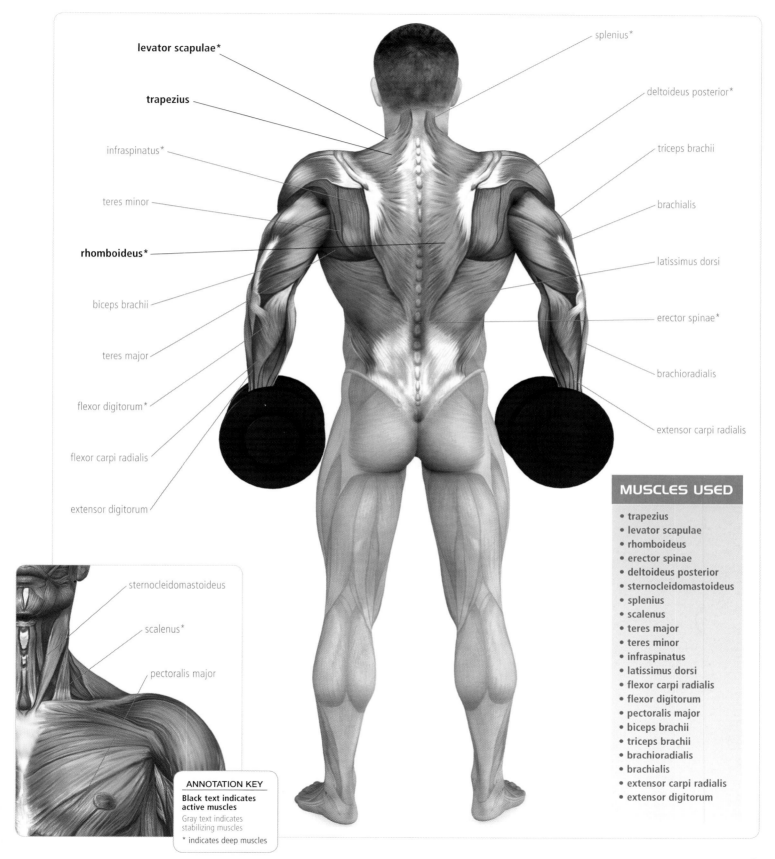

levator scapulae*

trapezius

infraspinatus*

teres minor

rhomboideus*

biceps brachii

teres major

flexor digitorum*

flexor carpi radialis

extensor digitorum

splenius*

deltoideus posterior*

triceps brachii

brachialis

latissimus dorsi

erector spinae*

brachioradialis

extensor carpi radialis

sternocleidomastoideus

scalenus*

pectoralis major

MUSCLES USED

- trapezius
- levator scapulae
- rhomboideus
- erector spinae
- deltoideus posterior
- sternocleidomastoideus
- splenius
- scalenus
- teres major
- teres minor
- infraspinatus
- latissimus dorsi
- flexor carpi radialis
- flexor digitorum
- pectoralis major
- biceps brachii
- triceps brachii
- brachioradialis
- brachialis
- extensor carpi radialis
- extensor digitorum

ANNOTATION KEY

Black text indicates active muscles

Gray text indicates stabilizing muscles

* indicates deep muscles

PLATE ROW

BACK

a

1 Standing with your feet parallel and placed slightly wider than shoulder width apart, grasp a weight plate with both hands.

2 Lean forward from the waist, bending your knees slightly.

TRAINER'S TIPS
• This is a great exercise for women.
• Try to stick out your chest and touch the flat side of the plate on each rep.
• Squeeze your arm muscles during the movement.
• Carefully control the speed of the weight's ascent and descent. Be careful not to drop the weight, as this can cause serious injury.

TARGET
• Upper back

LOOK FOR
• Your glutes and hamstrings to engage during the exercise.
• Your chin to remain up throughout the exercise.
• Your neck to remain long throughout the exercise.

AVOID
• Collapsing your chest.
• Rotating your shoulders inward.
• Using momentum to lift the weight.

3 Keeping your elbows in tight to your sides, extend your arms toward the floor.

4 Raise the weight back to the starting position, and repeat.

b

MUSCLES USED

• trapezius
• latissimus dorsi
• deltoideus posterior
• triceps brachii
• brachialis
• brachioradialis

• teres major
• erector spinae
• biceps brachii
• pectoralis major
• extensor digitorum
• rhomboideus

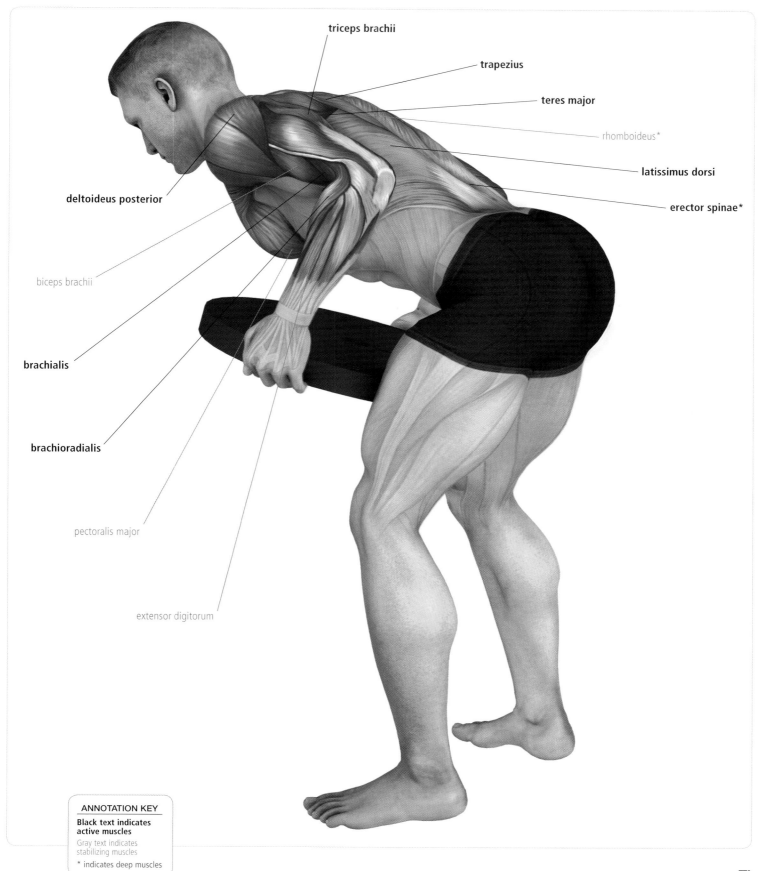

triceps brachii

trapezius

teres major

rhomboideus*

latissimus dorsi

erector spinae*

deltoideus posterior

biceps brachii

brachialis

brachioradialis

pectoralis major

extensor digitorum

ANNOTATION KEY

Black text indicates active muscles

Gray text indicates stabilizing muscles

* indicates deep muscles

DUMBBELL ROW

1 Holding a dumbbell in your left hand, stand next to an incline bench with your feet placed generously shoulder width apart.

a

2 Lean forward, and place your right hand on the bench. Your back should be flat and your knees slightly bent. Your left hand should be holding the dumbbell in a hammer-grip position, with your elbow close to your ribs.

b

3 Draw your elbow toward the ceiling.

4 Lower the dumbbell to starting position, and repeat. Switch sides, and repeat all steps with your right hand holding the dumbbell.

c

TRAINER'S TIPS
- Wear wrist straps for greater stability.
- Keep a slight bend in your supporting arm.
- Relax your jaw.

TARGET
- Middle back

LOOK FOR
- Your chest to remain up.
- Your pelvis to remain tucked slightly and your back flat.

AVOID
- Drawing your elbow away from your rib cage.
- Using momentum to lift the dumbbell.

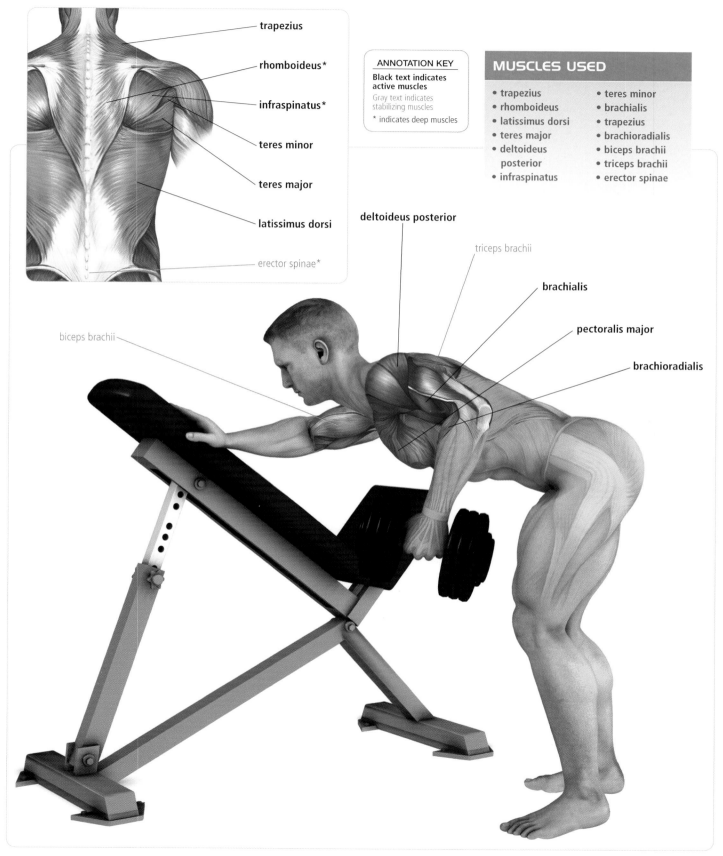

trapezius

rhomboideus*

infraspinatus*

teres minor

teres major

latissimus dorsi

erector spinae*

ANNOTATION KEY

Black text indicates active muscles

Gray text indicates stabilizing muscles

* indicates deep muscles

MUSCLES USED

- trapezius
- rhomboideus
- latissimus dorsi
- teres major
- deltoideus posterior
- infraspinatus
- teres minor
- brachialis
- trapezius
- brachioradialis
- biceps brachii
- triceps brachii
- erector spinae

deltoideus posterior

triceps brachii

brachialis

pectoralis major

brachioradialis

biceps brachii

CABLE SQUAT

1. Set a cable machine with a single-handle grip to its lowest setting.

2. Stand about four feet from the cable machine, with your feet parallel and shoulder width apart. Grasp a handle in each hand, with your palms facing inward in hammer-grip position.

3. Bend your knees and lower yourself into a seated squat position. This is your starting position.

4. Row the cable weight in toward your body, keeping your elbows close to your ribs.

5. Extend the weight back to the starting position, and repeat.

a

b

TARGETS
- Middle and lower back
- Gluteal muscles

LOOK FOR
- A constant engagement of the glutes and thighs while executing this exercise.

AVOID
- Collapsing your chest or rolling your shoulders in when extending the weight back to the starting position.

MUSCLES USED

- gluteus maximus
- rectus femoris
- vastus lateralis
- vastus intermedius
- vastus medialis
- adductor magnus
- soleus
- latissimus dorsi
- biceps femoris
- semitendinosus
- semimembranosus
- gastrocnemius

TRAINER'S TIP
- Exhale as you draw the weight in, and inhale as you extend your arms back to starting position.

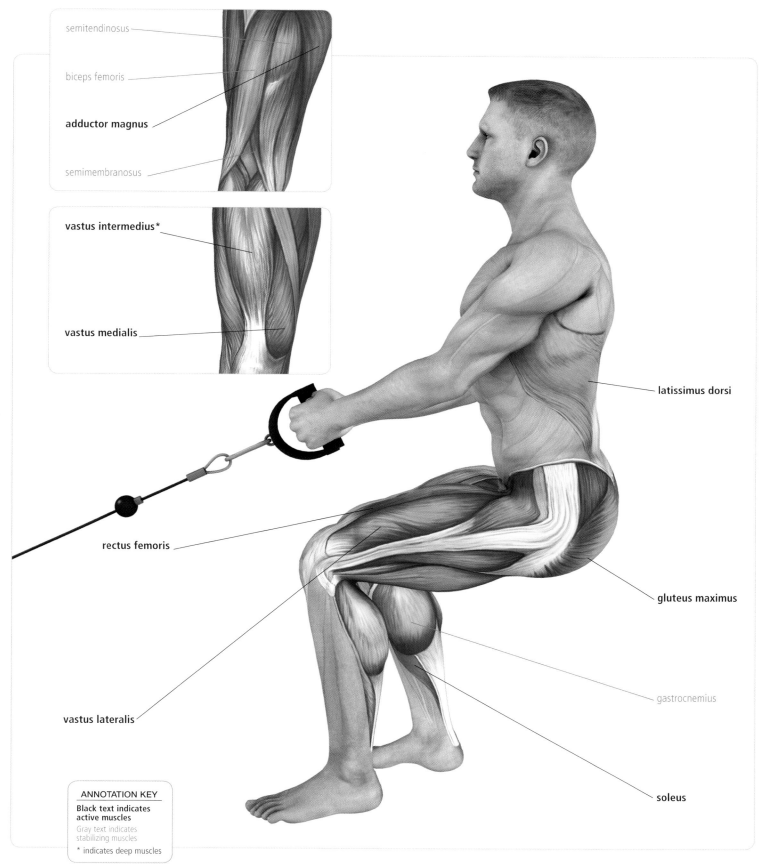

semitendinosus

biceps femoris

adductor magnus

semimembranosus

vastus intermedius*

vastus medialis

latissimus dorsi

rectus femoris

gluteus maximus

vastus lateralis

gastrocnemius

soleus

ANNOTATION KEY

Black text indicates active muscles

Gray text indicates stabilizing muscles

* indicates deep muscles

DUMBBELL DEADLIFT

① Standing with your feet placed shoulder width apart and your knees slightly bent, grasp a dumbbell in each hand, holding them by the sides of your body, palms facing each other. Keep your torso upright, and squeeze your shoulder blades together.

a

MUSCLES USED

- erector spinae
- biceps femoris
- semitendinosus
- gluteus maximus
- adductor magnus
- semimembranosus
- trapezius
- rhomboideus
- latissimus dorsi
- levator scapulae
- rectus abdominis
- obliquus externus
- obliquus internus

TARGET
- Lower back

LOOK FOR
- A constant spinal position throughout the movement.
- Your torso to flex forward from the hip.

AVOID
- Collapsing your chest and shoulders—keep your back flat.
- Allowing your shoulder blades to slip forward.

② Bend forward from the hips, keeping your knees stationary, and lower the dumbbells toward the tops of your feet. Rotate your palms slightly inward so that they face behind you as you lower the dumbbells until you feel a stretch in your hamstrings.

③ Return to the starting position, and repeat.

TRAINER'S TIPS
- This exercise is not recommended for anyone with a lower-back problem. Execute it with caution and care.
- When done correctly, the movement should feel like you are picking up something from the ground.
- Exhale as you lower the dumbbells, and inhale as you come up to the starting position.

b

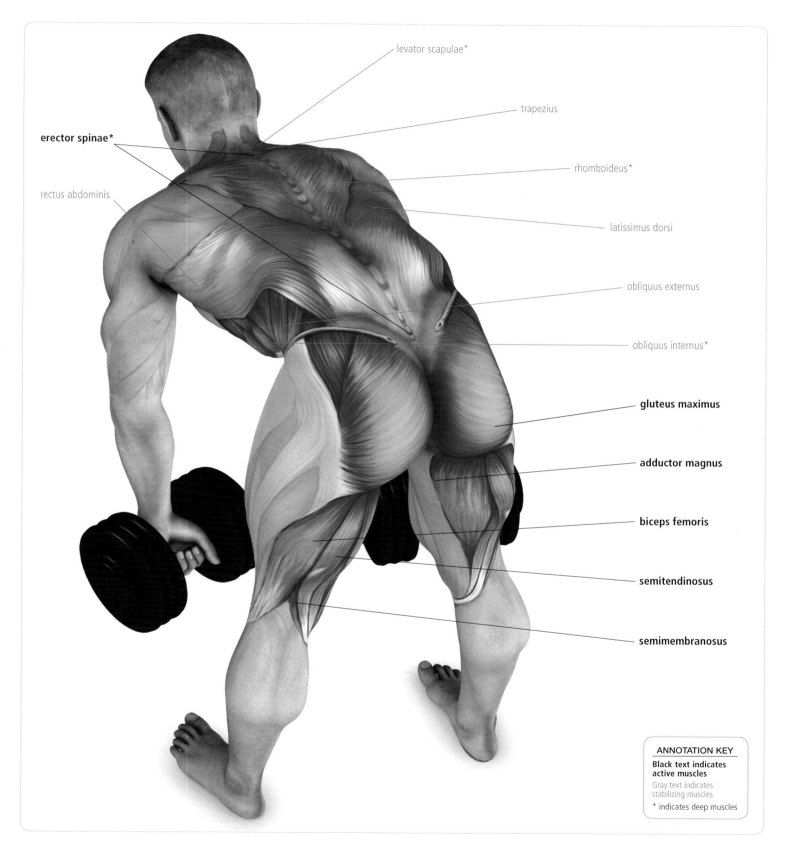

levator scapulae*

trapezius

erector spinae*

rhomboideus*

rectus abdominis

latissimus dorsi

obliquus externus

obliquus internus*

gluteus maximus

adductor magnus

biceps femoris

semitendinosus

semimembranosus

ANNOTATION KEY

Black text indicates active muscles
Gray text indicates stabilizing muscles
* indicates deep muscles

FLAT BENCH HYPEREXTENSIONS

1 Lie supine on a flat bench with your sternum even with the bench's upper edge. Your upper chest and head should hang off the top of the bench.

2 Hook your feet under the bench, securing yourself in a stable starting position. Place your hands at the sides of your head, fingertips touching your ears.

TRAINER'S TIPS
• Do not hold your breath during this exercise. Exhale as you raise your body, and inhale as you lower your body to starting position.
• Wear shoes that support a strong grip on the underside of the flat bench.

a

TARGET
• Lower back

LOOK FOR
• A constant engagement of the glutes and thighs while executing this exercise.
• Your lower body to remain taut throughout the movement.
• Your head to remain in neutral position.

AVOID
• Elevating your shoulders.
• Lifting your hip bones off the bench.

3 With arms bent and elbows out, raise your upper body about 8 to 12 inches off the bench.

4 Slowly and carefully lower your body to the starting position. Repeat.

MUSCLES USED
• **erector spinae**
• **gluteus maximus**
• **biceps femoris**
• **semitendinosus**
• **semimembranosus**
• **adductor magnus**
• **latissimus dorsi**
• **teres major**
• **deltoideus posterior**
• **triceps brachii**
• **brachialis**
• **brachioradialis**
• **biceps brachii**
• **trapezius**
• **pectoralis minor**
• **rhomboideus**
• **multifidus spinae**

b

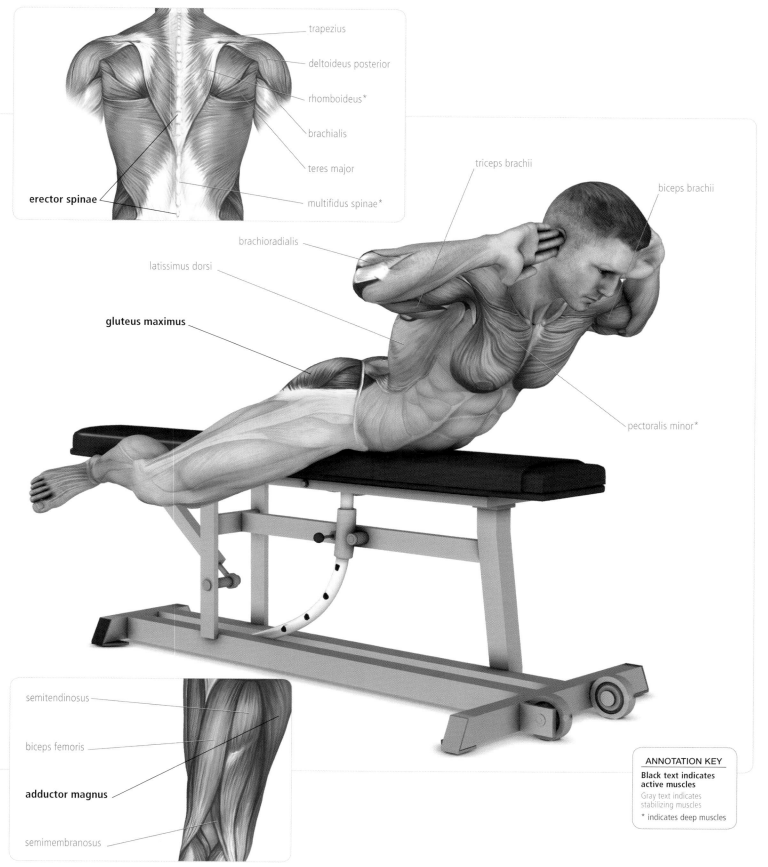

trapezius

deltoideus posterior

rhomboideus*

brachialis

teres major

erector spinae

multifidus spinae*

triceps brachii

biceps brachii

brachioradialis

latissimus dorsi

gluteus maximus

pectoralis minor*

semitendinosus

biceps femoris

adductor magnus

semimembranosus

ANNOTATION KEY

Black text indicates active muscles

Gray text indicates stabilizing muscles

* indicates deep muscles

GOOD MORNINGS

a

1. Set a barbell on a rack that best matches your height. Choose and rack the barbell with appropriate weights.

2. Holding onto the bar with both arms at each side, lift the barbell up and off the rack, using your legs and then straightening your torso.

3. Step forward, and stand with your feet parallel and shoulder width apart, keeping your head up to ensure a proper flat back.

TRAINER'S TIPS

• Inhale while lowering the torso, and exhale while raising the torso back to the starting position.
• This exercise should be performed inside a squat rack for safety.
• This is a very advanced exercise—execute with caution and care. If you are unsure of how heavy your weight should be, start with an extremely light weight.

b

TARGETS
• Lower back
• Hamstrings

LOOK FOR
• Your head to remain up throughout the movement.
• Your torso to move up and down through an arc of about 90 degrees.

AVOID
• Bending your torso below parallel to the floor.

4. Keeping your legs stationary, move your torso forward by bending at the hips. Slightly bending your knees, lower your torso until it is nearly parallel with the floor.

5. Raise your torso back to the starting position, and repeat.

MUSCLES USED

• **latissimus dorsi**
• **biceps femoris**
• **semitendinosus**
• **semimembranosus**
• **gluteus maximus**
• **adductor magnus**
• **erector spinae**
• **multifidus spinae**
• **rectus abdominis**
• **obliquus externus**
• **obliquus internus**

80

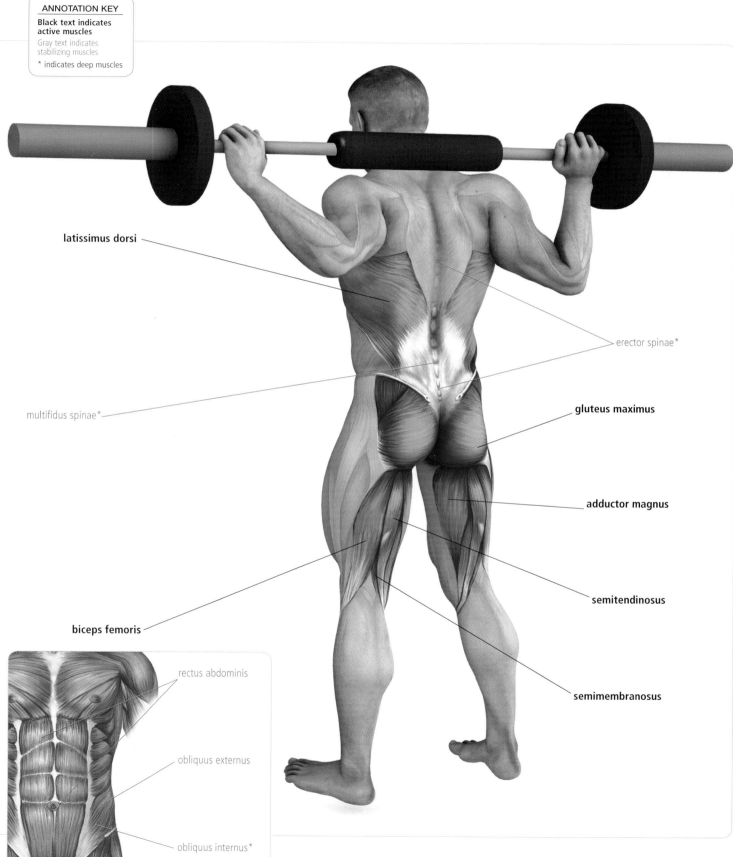

ANNOTATION KEY
**Black text indicates
active muscles**
Gray text indicates
stabilizing muscles
* indicates deep muscles

latissimus dorsi

erector spinae*

multifidus spinae*

gluteus maximus

adductor magnus

semitendinosus

biceps femoris

semimembranosus

rectus abdominis

obliquus externus

obliquus internus*

SHOULDERS

The shoulder joint has the greatest range of motion of any joint in the human body. The deltoid muscle, divided into front, side, and back sections called the deltoideus anterior, medialis, and posterior, forms the outer layer of shoulder muscle. Known as the "delts," these are the largest and strongest muscles of the shoulder. The delts are responsible for elevating and rotating the arm. The trapezius, located on the back, also works to enable shoulder movement, elevating and retracting the shoulder blades.

Along with the deltoids, are a group of muscles that stabilize the shoulder, known collectively as the rotator cuff. The rotator cuff is made up of the infraspinatus, subscapularis, supraspinatus, and teres minor. Rotator cuff injuries (usually due to bad workout form) cause many gym-goers to have underdeveloped shoulder muscles.

Careful targeting of this group of muscles results in strong, developed shoulders that give the appearance of a broad, worked-out body.

BARBELL ROW

❶ Stand with your feet parallel and shoulder width apart, your knees slightly bent.

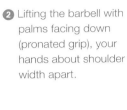

❷ Lifting the barbell with palms facing down (pronated grip), your hands about shoulder width apart.

b

a

❸ Bend at the waist to bring your torso forward, keeping a straight back until it is nearly parallel to the floor. The barbell should be directly in front of you, allowing your arms to hang perpendicular to the floor and your torso. This is the starting position.

❹ Lift the barbell toward your torso, keeping your elbows pointing in toward the sides of the body.

c

TARGETS
• Deltoids
• Back

LOOK FOR
• Your torso to remain horizontal throughout the exercise.

AVOID
• Dropping your head during this exercise.

❺ Slowly lower the weight to the starting position. Repeat.

d

MUSCLES USED

• deltoideus posterior	• triceps brachii
• trapezius	• erector spinae
• rhomboideus	• gluteus maximus
• latissimus dorsi	• biceps femoris
• teres major	• semitendinosus
• infraspinatus	• semimembranosus
• brachialis	• adductor magnus
• brachioradialis	• obliquus externus
• pectoralis major	• obliquus internus
• biceps brachii	• rectus abdominis

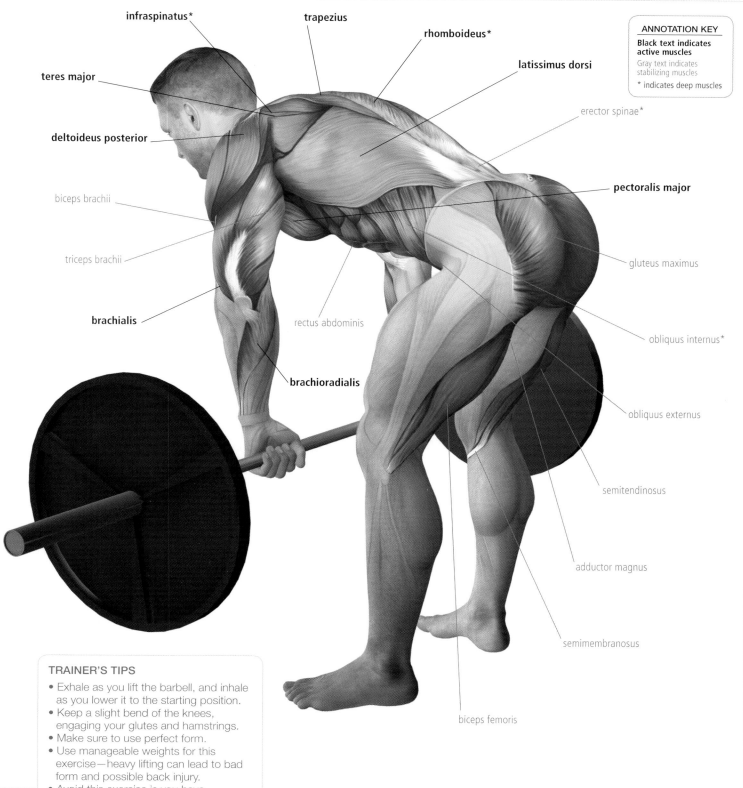

infraspinatus*

trapezius

rhomboideus*

latissimus dorsi

ANNOTATION KEY

Black text indicates active muscles

Gray text indicates stabilizing muscles

* indicates deep muscles

erector spinae*

teres major

deltoideus posterior

biceps brachii

triceps brachii

brachialis

brachioradialis

rectus abdominis

pectoralis major

gluteus maximus

obliquus internus*

obliquus externus

semitendinosus

adductor magnus

semimembranosus

biceps femoris

TRAINER'S TIPS

- Exhale as you lift the barbell, and inhale as you lower it to the starting position.
- Keep a slight bend of the knees, engaging your glutes and hamstrings.
- Make sure to use perfect form.
- Use manageable weights for this exercise—heavy lifting can lead to bad form and possible back injury.
- Avoid this exercise is you have back problems.

PYRAMID CABLE PRESS

1 Set both sides of a cable machine to the lowest settings. Attach a single-handle grip to each side.

2 Center yourself between the cable uprights, with your feet shoulder width apart and your pelvis tucked in. Pick up the handle grips one at a time.

a

3 Starting with your palms facing up, curl the weight in while simultaneously bending your knees. Then rotate your palms to face forward, with your arms in a 90-degree bend to start.

b

4 Press the cable weight upward in a pyramid motion to a near touch at the top range of the movement.

5 Slowly and deliberately lower the weight to the starting position. Repeat.

c

TARGET
• Deltoids

LOOK FOR
• A consistent speed of movement throughout the execution of this exercise.

AVOID
• Lifting weights that are too heavy—this will put great strain on the biceps as you get into the starting position and may cause injury.
• Bending your knees to give assistance and momentum while executing this movement.

MUSCLES USED

• **deltoideus anterior**
• **deltoideus medialis**
• **supraspinatus**
• **triceps brachii**
• **trapezius**
• **serratus anterior**
• **biceps brachii**
• **levator scapulae**

TRAINER'S TIP
• Exhale as you press the cable weight up, and inhale as you lower it down to the starting position.

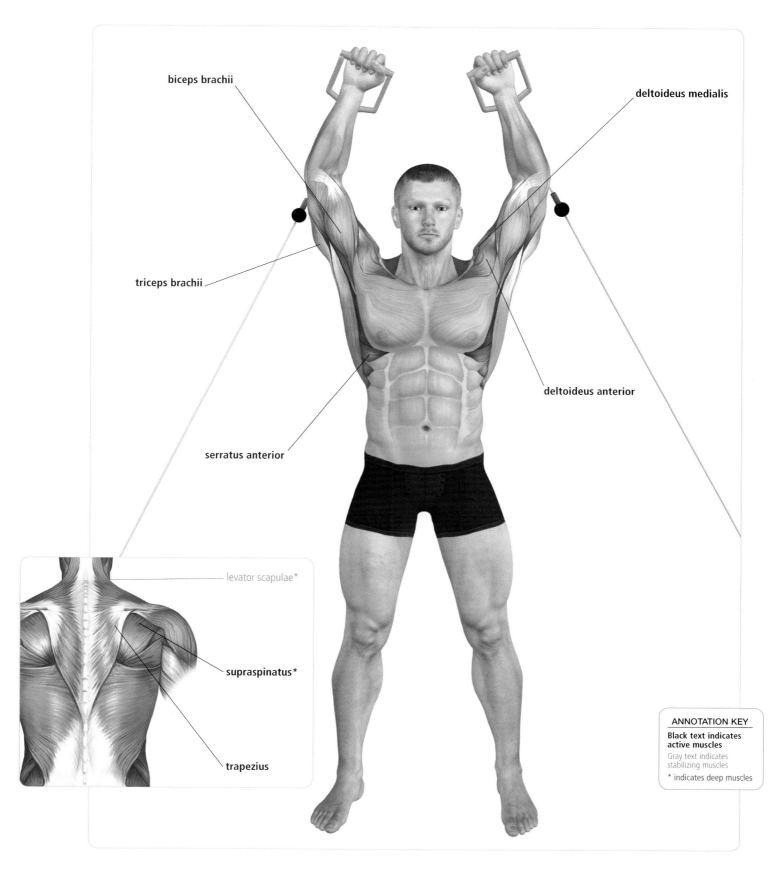

biceps brachii

deltoideus medialis

triceps brachii

deltoideus anterior

serratus anterior

levator scapulae*

supraspinatus*

trapezius

ANNOTATION KEY

Black text indicates active muscles

Gray text indicates stabilizing muscles

* indicates deep muscles

FRONT PLATE RAISE

1 Holding onto a 45-pound plate with both hands in hammer-grip position, stand with your feet parallel and shoulder width apart and your pelvis tucked in slightly.

2 Raise the plate to shoulder height.

3 Slowly lower the weight back to the starting position.

TARGET
• Deltoids

LOOK FOR
• A steady, controlled movement.

AVOID
• Hyperextending your elbows while lifting the weight.
• Allowing your shoulders to rotate inward.

MUSCLES USED

• deltoideus anterior
• deltoideus posterior
• deltoideus medialis
• trapezius
• serratus anterior
• levator scapulae
• brachialis
• biceps brachii
• brachioradialis
• flexor digitorum
• flexor carpi radialis

TRAINER'S TIPS
• Exhale as you lift the plate, and inhale as you lower the plate.
• Keep your posture erect while executing the exercise.
• Keep your shoulders down and back away from your ears.

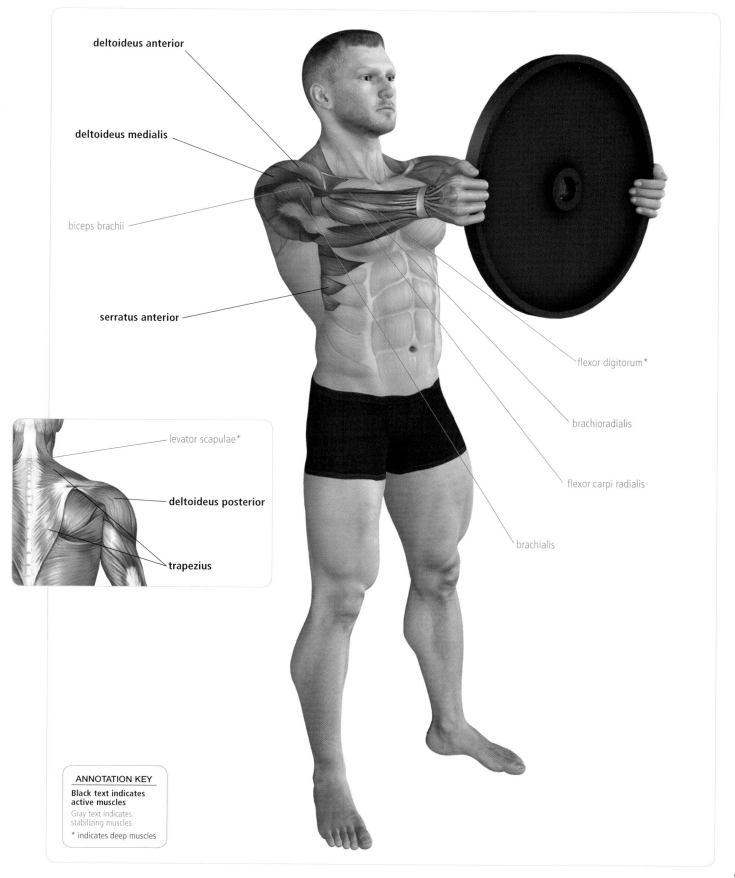

deltoideus anterior

deltoideus medialis

biceps brachii

serratus anterior

flexor digitorum*

brachioradialis

flexor carpi radialis

brachialis

levator scapulae*

deltoideus posterior

trapezius

ANNOTATION KEY

Black text indicates active muscles

Gray text indicates stabilizing muscles

* indicates deep muscles

SEATED ARNOLD PRESS

① Lie back on an incline bench, and place a pair of dumbbells on your thighs to start. Kick up dumbbells with your elbows in and lift them to shoulder height.

a

② Raise the dumbbells as you rotate the palms of your hands until they are facing forward.

b

c

③ Press upward until your arms are extended above your head.

④ After a slight pause at the top, slowly lower the dumbbells back to the starting position while rotating the palms of your hands toward you. Repeat.

TARGETS
- Front and side deltoids

LOOK FOR
- The starting position to look like the contracted portion of a dumbbell curl.

AVOID
- Hyperextending your back while pressing the dumbbells.

TRAINER'S TIPS
- Exhale as you press the dumbbells up, and inhale as you lower them back to the starting position.
- Keep your chest elevated and your shoulders down and back away from your ears.
- Performing this exercise while seated prevents you from using upward momentum to lift the dumbbells.

levator scapulae*

deltoideus medialis

deltoideus posterior

supraspinatus*

trapezius

deltoideus anterior

triceps brachii

serratus anterior

ANNOTATION KEY

Black text indicates active muscles

Gray text indicates stabilizing muscles

* indicates deep muscles

MUSCLES USED

- deltoideus anterior
- deltoideus medialis
- deltoideus posterior
- supraspinatus
- triceps brachii
- serratus anterior
- levator scapulae
- trapezius

DUMBBELL SHOULDER PRESS

1 Lie back on an incline bench, and place a pair of dumbbells on your thighs to start. Kick up dumbbells with your elbows in, and then lift them to shoulder height.

2 Turn your elbows out in a 90-degree angle of the arms, with your palms facing forward.

MUSCLES USED

- deltoideus anterior
- deltoideus medialis
- supraspinatus
- triceps brachii
- trapezius
- serratus anterior
- pectoralis major
- biceps brachii
- levator scapulae

3 Press the dumbbells upward into a pyramid position.

4 Slowly lower the dumbbells back to the starting position. Repeat.

TARGET
- Side deltoids

LOOK FOR
- Your chin to remain over your shoulders during this exercise.

AVOID
- Hyperextending your back when pressing the dumbbells upward during the press motion of this exercise.

TRAINER'S TIPS
- When you are finished with the set, bring your elbows back in, palms facing each other, and slowly lower the dumbbells back to your thighs.
- Exhale as you press the dumbbells up, and inhale as you lower them down.
- Relax your neck and jaw during the exercise.

deltoideus medialis

deltoideus anterior

biceps brachii

triceps brachii

pectoralis major

serratus anterior

levator scapulae

supraspinatus*

trapezius

ANNOTATION KEY

Black text indicates active muscles

Gray text indicates stabilizing muscles

* indicates deep muscles

STANDING BARBELL ROW

1 Grasp a barbell with both hands, palms facing down in an overhead grip about shoulder width apart or slightly narrower. Stand with your feet parallel and shoulder width apart, your knees slightly bent, and your pelvis tucked in slightly.

2 To get into starting position, rest the barbell against the top of your thighs. Your arms should be extended with a slight bend in at the elbows. Your back should be straight.

3 Focus on engaging your side deltoids to lift the barbell to chest height. Pause at the top of the movement, and return to starting position. Repeat.

MUSCLES USED

- deltoideus anterior
- supraspinatus
- brachialis
- brachioradialis
- biceps brachii
- trapezius
- serratus anterior
- teres minor
- levator scapulae

TARGET
- Side deltoids

LOOK FOR
- Your elbows to initiate the movement.
- Your elbows to raise higher than your forearms.
- Your torso to remain stationary.

AVOID
- Lifting weights that are too heavy with this exercise— excessive weight can lead to bad form and a possible shoulder injury.

TRAINER'S TIPS
- The barbell should be close to the body as you move it up.
- Exhale as you lift the barbell, and inhale as you lower it to the starting position.

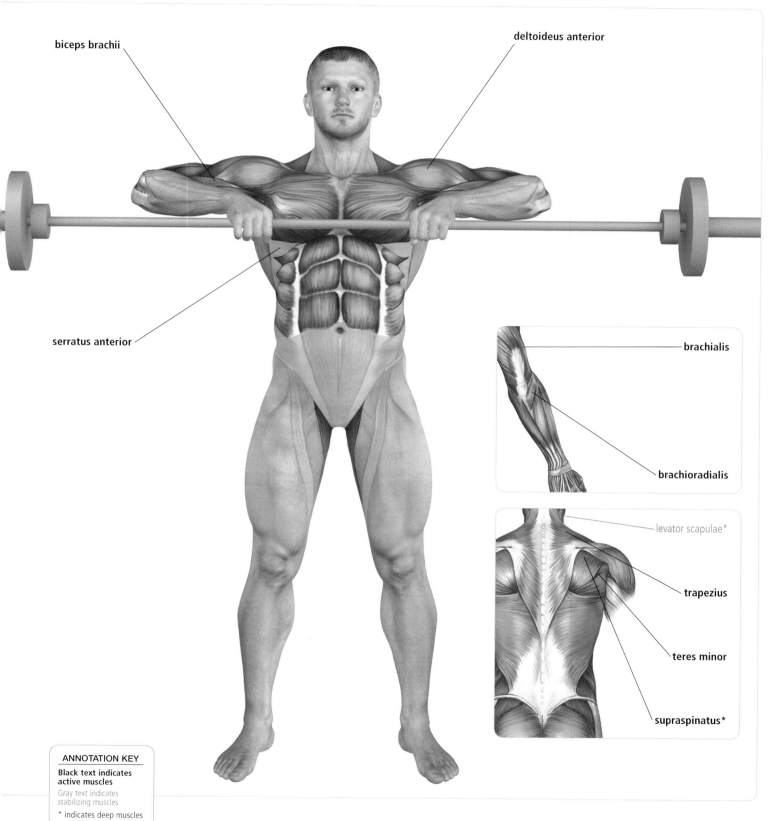

biceps brachii

deltoideus anterior

serratus anterior

brachialis

brachioradialis

levator scapulae*

trapezius

teres minor

supraspinatus*

ANNOTATION KEY

Black text indicates active muscles

Gray text indicates stabilizing muscles

* indicates deep muscles

DUMBBELL SIDE PUSH-OUT

a

1 Holding a dumbbell in each hand, stand with your feet parallel and shoulder width apart, your knees slightly bent and your pelvis slightly tucked. Your elbows should be slightly bent, your chest elevated, and your shoulders down and back away from your ears.

b

2 With palms facing in toward your body, push out about four inches away from the body with the back of your hands.

3 Slowly bring the dumbbells back to the starting position. Repeat.

TARGET
- Side deltoids

LOOK FOR
- A steady, controlled movement.

AVOID
- Rushing the exercise.

TRAINER'S TIPS
- Relax your neck and jaw.
- Exhale as you push the dumbbells out away from your body, and inhale as you bring them back to the starting position.
- Wear wrist straps to stabilize your movements.

96

trapezius

deltoideus anterior

deltoideus medialis

biceps brachii

triceps brachii

serratus anterior

MUSCLES USED

- deltoideus medialis
- deltoideus anterior
- triceps brachii
- biceps brachii
- trapezius
- serratus anterior

ANNOTATION KEY

Black text indicates active muscles

Gray text indicates stabilizing muscles

* indicates deep muscles

LATERAL SHOULDER RAISE

a

1 Holding a dumbbell in each hand, stand with your feet parallel and shoulder width apart, your knees slightly bent. Bend your elbow slightly and face your palms in toward the body.

MUSCLES USED

- deltoideus medialis
- deltoideus anterior
- suprapinatus
- trapezius
- serratus anterior
- pectoralis minor
- teres minor
- infrapinatus
- rhomboideus
- erector spinae
- levator scapulae
- triceps brachii
- brachialis
- biceps brachii
- brachioradialis
- flexor digitorum
- flexor carpi radialis
- latissimus dorsi
- pectoralis major

2 Extend both arms out to the sides to shoulder height.

3 Slowly lower the dumbbells back to the starting position. Repeat.

TARGET
- Side deltoids

LOOK FOR
- Your elbows to remain in a fixed and slightly bent position throughout the movement.
- Your elbows to be directly lateral to your shoulders at the top of the movement.

AVOID
- Using momentum to lift the dumbbells.

b

TRAINER'S TIPS
- Exhale as you lift the dumbbells, and inhale as you lower them.
- Keep your chest elevated and your shoulders down and back away from your ears.
- Do not allow your elbows to drop lower than your wrists—this will make the front deltoids the primary movers instead of the lateral deltoids.

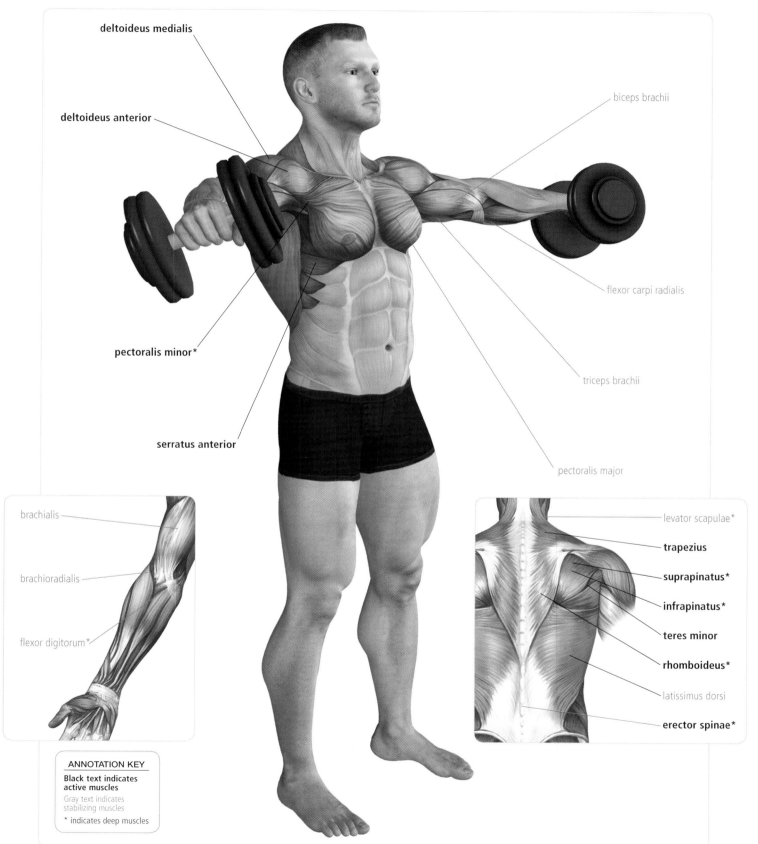

deltoideus medialis

deltoideus anterior

biceps brachii

pectoralis minor*

flexor carpi radialis

triceps brachii

serratus anterior

pectoralis major

brachialis

brachioradialis

flexor digitorum*

levator scapulae*

trapezius

suprapinatus*

infrapinatus*

teres minor

rhomboideus*

latissimus dorsi

erector spinae*

ANNOTATION KEY

Black text indicates active muscles

Gray text indicates stabilizing muscles

* indicates deep muscles

BENT-OVER CABLE RAISE

1 Set a cable machine with a the single-handle grip to its lowest setting.

2 Stand with your right shoulder parallel to the cable machine.

3 With your left hand, grasp the handle with your palm facing inward in the hammer-grip position.

4 With your feet parallel and shoulder width apart, slightly bend your knees, as you lean over until your back is nearly flat.

a

TRAINER'S TIPS
- Aim for getting a stretch in the rear deltoid during the starting position.
- Make sure to engage the glutes and thighs during this exercise.

5 Extend your left arm out to the side.

6 Carefully bring the cable weight back to the starting position, and repeat. Switch sides, and repeat all steps with your right hand holding the handle grip.

TARGETS
- Side and rear deltoids

LOOK FOR
- Your shoulders to remain down and back.

AVOID
- Using momentum to execute the movement.
- Dropping your head.
- Allowing the chest or shoulders to roll in during the exercise.

b

MUSCLES USED

- deltoideus medialis
- deltoideus posterior
- deltoideus anterior
- triceps brachii
- pectoralis major
- serratus anterior
- biceps brachii
- levator scapulae
- trapezius

trapezius

levator scapulae*

deltoideus medialis

deltoideus posterior

triceps brachii

deltoideus anterior

biceps brachii

pectoralis major

serratus anterior

ANNOTATION KEY

Black text indicates active muscles

Gray text indicates stabilizing muscles

* indicates deep muscles

ONE-ARM DUMBBELL RAISE

a

1. Stand next to an incline bench with a dumbbell in your right hand. Place your left hand on the back of the bench for balance and support.

2. With your feet parallel and shoulder width apart and your knees slightly bent, lean forward with your back flat.

MUSCLES USED

- deltoideus posterior
- deltoideus medialis
- infraspinatus
- teres minor
- trapezius
- rhomboideus
- brachialis
- biceps brachii
- brachioradialis
- flexor digitorum
- flexor carpi radialis
- latissimus dorsi
- pectoralis major
- gluteus maximus
- adductor magnus

3. Extend your right arm out to the side, lifting the dumbbell in line with your shoulder.

4. Slowly lower the dumbbell back to the starting position, and repeat. Switch sides, and repeat all steps with your left hand holding the dumbbell.

TARGETS
- Rear deltoids

LOOK FOR
- Your glutes and thighs to fully engage while executing this exercise.

AVOID
- Rolling your chest or shoulders forward and in while executing this exercise.

TRAINER'S TIPS
- Exhale as you lift the dumbbell, and inhale as you bring them back to the starting position.
- Keep your chin up.

b

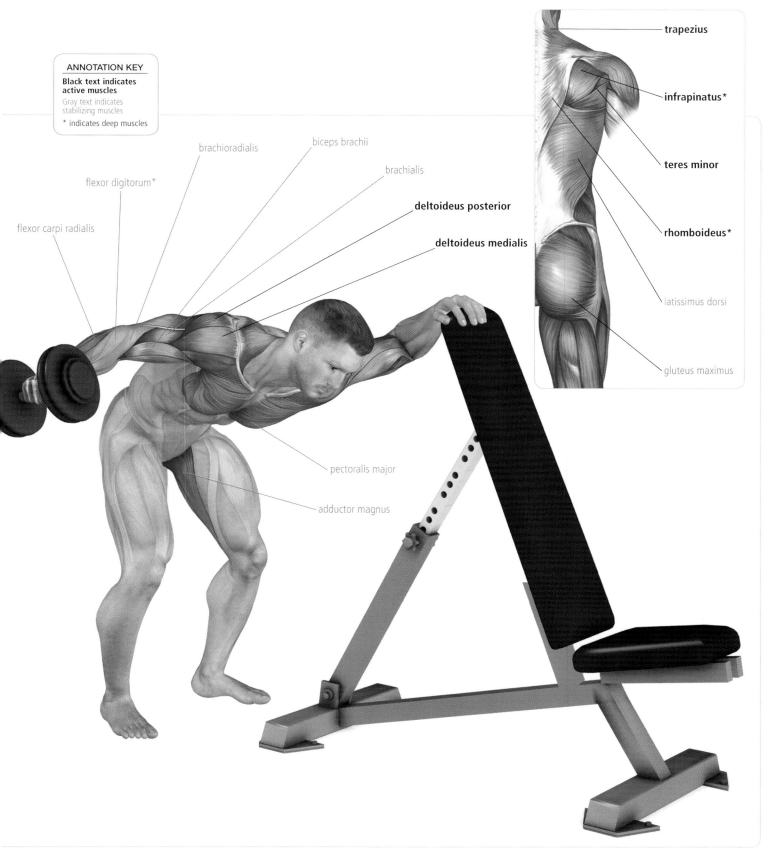

trapezius

infrapinatus*

teres minor

rhomboideus*

latissimus dorsi

gluteus maximus

ANNOTATION KEY

Black text indicates active muscles

Gray text indicates stabilizing muscles

* indicates deep muscles

brachioradialis

biceps brachii

brachialis

flexor digitorum*

deltoideus posterior

flexor carpi radialis

deltoideus medialis

pectoralis major

adductor magnus

REVERSE FLY

a

1 Holding a dumbbell in each hand straddle an incline bench, facing backward. With your hands in the hammer-grip position, lower the dumbbells off the incline bench.

2 Lower your body to the bench as you simultaneously lower the dumbbells to the starting position.

MUSCLES USED

- deltoideus anterior
- deltoideus posterior
- deltoideus medialis
- trapezius
- levator scapulae
- brachialis
- biceps brachii
- brachioradialis
- flexor digitorum
- flexor carpi radialis

3 With you palms facing in toward each other, draw your arms up to the side and away from the body.

b

TARGET
- Rear deltoids

LOOK FOR
- A steady, controlled movement during both the ascent and descent phases of the exercise.

AVOID
- Holding onto neck and jaw tension.
- Sliding down the bench while executing this exercise.

4 Lift until you reach shoulder height, and then lower the dumbbells back to the starting position. Repeat.

TRAINER'S TIPS
- Keep your feet firmly planted on floor.
- Keep your chest elevated while executing this exercise.
- Exhale as you lift the dumbbells up, and inhale as you bring them back to the starting position.

c

deltoideus posterior

deltoideus medialis

trapezius

levator scapulae*

deltoideus anterior

biceps brachii

brachialis

brachioradialis

flexor carpi radialis

flexor digitorum*

ANNOTATION KEY
Black text indicates active muscles
Gray text indicates stabilizing muscles
* indicates deep muscles

ARMS

The arm muscles are diverse and numerous, but the main focus of those who want to build their muscles are the biceps brachii and the triceps brachii. The biceps, located on the upper arm, consists of two bundles of muscles or "heads." It is a flexor muscle, meaning that it bends the lower arm toward the upper arm. The triceps is a three-headed extensor muscle. It works in opposition to the biceps to extend the lower arm and straighten the elbow.

Other major arm muscles include the brachialis, brachioradialis, coracobrachialis, pronator teres, and palmaris longus, along with the wrist flexors (the flexor carpi radialis and flexor carpi ulnaris) and the wrist extensors (the extensor carpi radialis and extensor carpi ulnaris).

TRICEPS BENCH DIP

1 Using either two flat benches, or a flat bench and a step-up stool, sit on the bench and grasp the edge of bench. Carefully place your feet shoulder width apart on the opposite bench or stool.

MUSCLES USED

- triceps brachii
- deltoideus anterior
- pectoralis major
- rhomboideus
- levator scapulae
- latissimus dorsi
- biceps brachii
- trapezius

a

2 With your elbows pointing straight behind you, lower your glutes about four to eight inches.

3 Focus on engaging the triceps to lift your body back to the starting position. Repeat.

TRAINER'S TIPS
- Inhale as you lower your body, and exhale as you lift back to the starting position.
- When you are finished with this exercise, carefully place your glutes back on the bench, and slowly lower your feet to the floor.

b

TARGET
- Triceps

LOOK FOR
- Your chest to remain elevated throughout the exercise.
- Proper head placement—your chin should be slightly elevated.

AVOID
- Lowering your body too far below 90 degrees—you risk straining your shoulders or injuring your rotator cuff.

ANNOTATION KEY

Black text indicates active muscles

Gray text indicates stabilizing muscles

* indicates deep muscles

levator scapulae*

rhomboideus*

trapezius

biceps brachii

deltoideus anterior

triceps brachii

pectoralis major

latissimus dorsi

SKULL CRUSHERS

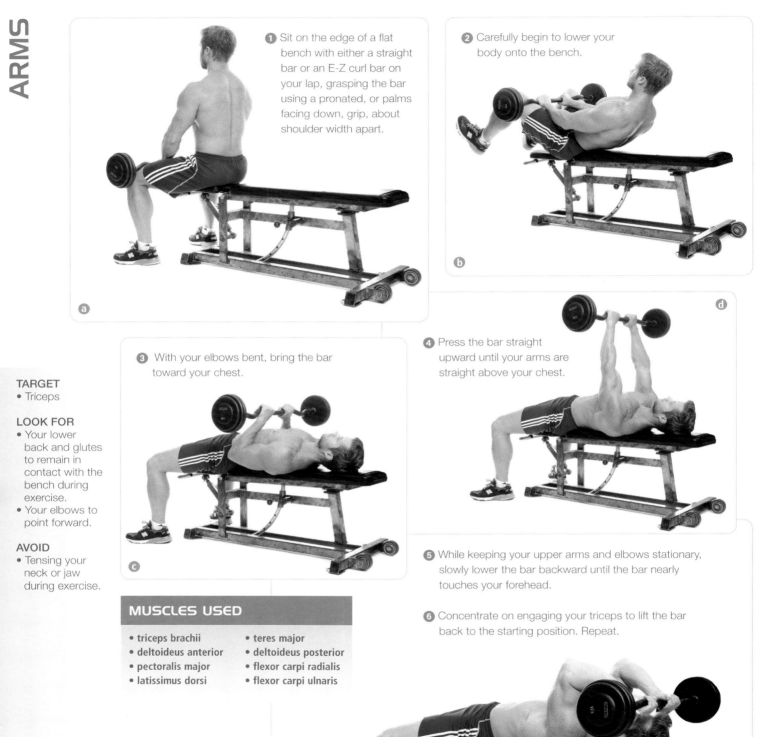

1 Sit on the edge of a flat bench with either a straight bar or an E-Z curl bar on your lap, grasping the bar using a pronated, or palms facing down, grip, about shoulder width apart.

2 Carefully begin to lower your body onto the bench.

3 With your elbows bent, bring the bar toward your chest.

4 Press the bar straight upward until your arms are straight above your chest.

5 While keeping your upper arms and elbows stationary, slowly lower the bar backward until the bar nearly touches your forehead.

6 Concentrate on engaging your triceps to lift the bar back to the starting position. Repeat.

TARGET
• Triceps

LOOK FOR
• Your lower back and glutes to remain in contact with the bench during exercise.
• Your elbows to point forward.

AVOID
• Tensing your neck or jaw during exercise.

MUSCLES USED

• triceps brachii	• teres major
• deltoideus anterior	• deltoideus posterior
• pectoralis major	• flexor carpi radialis
• latissimus dorsi	• flexor carpi ulnaris

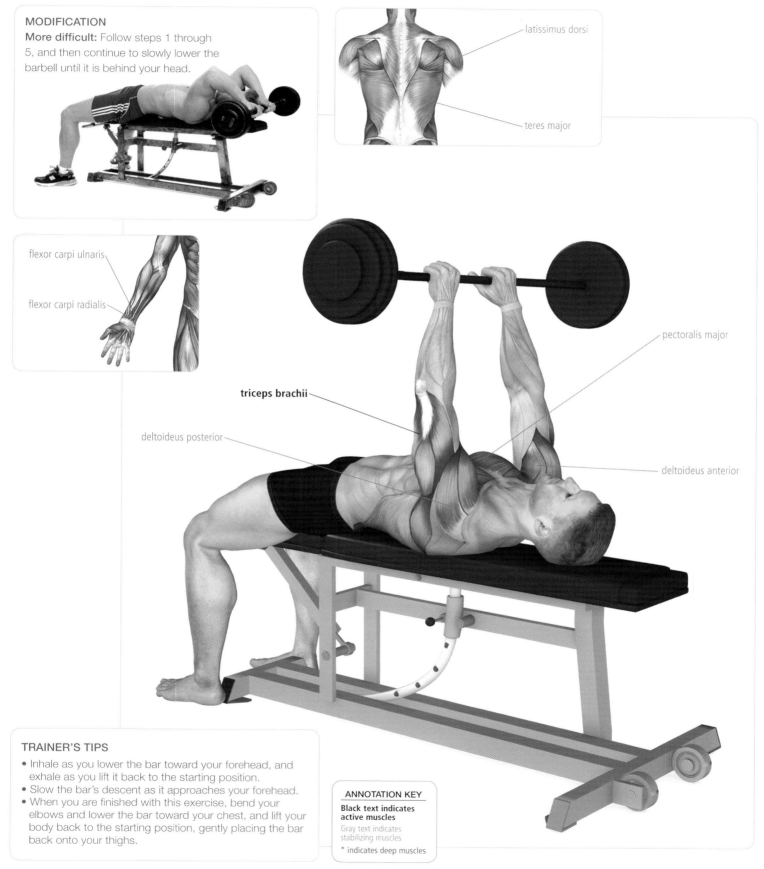

MODIFICATION

More difficult: Follow steps 1 through 5, and then continue to slowly lower the barbell until it is behind your head.

latissimus dorsi

teres major

flexor carpi ulnaris

flexor carpi radialis

triceps brachii

deltoideus posterior

pectoralis major

deltoideus anterior

TRAINER'S TIPS

- Inhale as you lower the bar toward your forehead, and exhale as you lift it back to the starting position.
- Slow the bar's descent as it approaches your forehead.
- When you are finished with this exercise, bend your elbows and lower the bar toward your chest, and lift your body back to the starting position, gently placing the bar back onto your thighs.

ANNOTATION KEY

Black text indicates active muscles

Gray text indicates stabilizing muscles

* indicates deep muscles

SMITH MACHINE CLOSE-GRIP PRESS

ARMS

① Lie supine on the flat bench of a Smith machine. Grasp the barbell with your hands about 8 to 10 inches apart. Lift and unhook the barbell.

a

MUSCLES USED

- triceps brachii
- deltoideus anterior
- pectoralis major
- biceps brachii

② Keeping your elbows close to your ribs, lower the bar until it nearly touches your chest. Lift the bar back toward the starting position, and repeat.

TARGET
- Triceps

LOOK FOR
- Your elbows to remain tucked toward the sides of your body throughout the exercise.
- Your feet to remain grounded on the floor.

AVOID
- Lifting your glutes and lower back off the bench.

b

TRAINER'S TIPS
- Inhale as you lower the barbell, and exhale as you raise it.
- Start with a light weight until you feel secure and confident with this exercise.
- When you are finished with this exercise, carefully hook the barbell back onto the machine.

deltoideus anterior

pectoralis major

triceps brachii

biceps brachii

ANNOTATION KEY

Black text indicates active muscles

Gray text indicates stabilizing muscles

* indicates deep muscles

ONE-ARM CABLE EXTENSION

1 Set a cable machine with a single-handle grip to its highest setting. Stand facing the cable machine upright, with your back straight and feet parallel and about shoulder width apart.

2 Grasp the handle with your left hand, palm facing upward to start.

a

TRAINER'S TIPS
- For proper position, keep your neck back and your chin slightly up.
- Exhale as you lower the cable weight, and inhale as you return it to the starting position.

b

TARGET
- Triceps

LOOK FOR
- Your working elbow to remain close to the side of your body.
- Your shoulders to remain down and back, away from ears.

AVOID
- Using momentum to execute movement—concentrate on isolating and using the triceps muscle.

3 Lower the cable weight toward the side of your left thigh.

4 Return the cable weight to the starting position, and repeat. Switch sides, and repeat all steps with your right hand holding the handle grip.

MUSCLES USED

- triceps brachii
- latissimus dorsi
- teres major
- deltoideus posterior
- pectoralis major
- pectoralis minor
- trapezius
- obliquus externus
- obliquus internus
- rectus abdominis
- extensor carpi radialis
- extensor carpi ulnaris

MODIFICATION

Similar difficulty: Follow steps 1 and 2 of the One-Arm Cable Extension exercise, but while lowering the cable weight, turn your palm so that it faces your thigh (about a quarter turn of the wrist). Then while raising the weight back to the starting position, turn your palm back so that it again faces upward.

MODIFICATION

Similar difficulty: Follow step 1 of the One-Arm Cable Extension exercise, but grasp the handle-grip with your palm facing downward. Lower the cable weight toward the side of your thigh, keeping the palm facing down. Return the cable weight to the starting position.

ANNOTATION KEY

Black text indicates active muscles

Gray text indicates stabilizing muscles

* indicates deep muscles

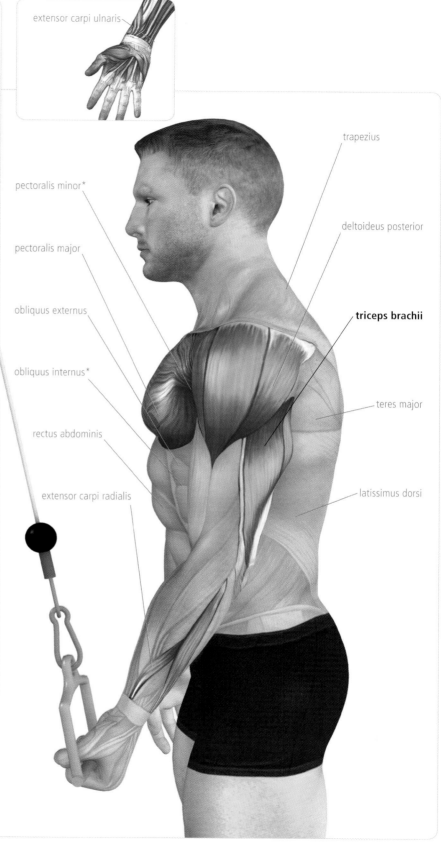

extensor carpi ulnaris

trapezius

pectoralis minor*

deltoideus posterior

pectoralis major

triceps brachii

obliquus externus

obliquus internus*

teres major

rectus abdominis

latissimus dorsi

extensor carpi radialis

ROPE OVERHEAD EXTENSION

1 Attach a rope attachment to the lowest setting of a cable machine. Grasp the rope with both hands, keeping your elbows close to the body.

2 Maintaining a wide stance, begin to pull upward on the rope.

3 Continue to turn forward as you slide your hands up above your head.

TARGET
• Triceps

LOOK FOR
• Your upper arms to remain stationary as you lower the weight behind your head.

AVOID
• Using momentum to execute movement— concentrate on isolating and using the triceps muscle.

4 Your starting position will be facing forward, with your elbows close to your head and your arms perpendicular to the floor with your knuckles pointed towards the ceiling.

5 Slowly lower the weight behind your head.

MUSCLES USED

• triceps brachii
• latissimus dorsi
• teres major
• deltoideus posterior
• pectoralis major
• pectoralis minor
• trapezius
• obliquus externus
• obliquus internus
• rectus abdominis
• flexor carpi radialis
• flexor carpi ulnaris

flexor carpi ulnaris

flexor carpi radialis

6 Raise the cable weight back to the starting position, and repeat.

triceps brachii

pectoralis minor*

pectoralis major

rectus abdominis

TRAINER'S TIPS

- Inhale as you lower the weight, and exhale as you lift the weight back to the starting position.
- Take a short pause when your triceps are fully stretched at the bottom of the range of movement.
- When you are finished with this exercise, slowly and carefully lower the rope, keeping your elbows close the body to avoid shoulder injury.

latissimus dorsi

obliquus externus

obliquus internus*

trapezius

deltoideus posterior

teres major

ANNOTATION KEY

Black text indicates active muscles

Gray text indicates stabilizing muscles

* indicates deep muscles

ROPE PUSH-DOWN

1. Attach a rope attachment to the highest setting of a cable machine.

2. Stand with your feet parallel and shoulder width, your knees slightly bent, and your pelvis tucked in slightly. Grasp the rope with both hands in hammer-grip position.

3. Keeping your elbows tucked in toward the sides of your body, lower the weight down toward your thighs.

4. Slowly raise the cable weight back to the starting position.

TARGET
- Triceps

LOOK FOR
- Your upper arms to remain stationary throughout the exercise.
- Your wrists to stay in line with your forearms.

AVOID
- Bending your wrists while lowering the weight.
- Using momentum to execute movement—concentrate on isolating and using the triceps muscle.

MUSCLES USED

- triceps brachii
- latissimus dorsi
- teres major
- deltoideus posterior
- pectoralis major
- pectoralis minor
- trapezius
- obliquus externus
- obliquus internus
- rectus abdominis
- flexor carpi radialis
- extensor carpi ulnaris

TRAINER'S TIPS
- For proper position, keep your neck back and your chin slightly up.
- Exhale as you lower the cable weight, and inhale as you bring it back to the starting position.
- Take a slight pause at the lowest point of the movement.

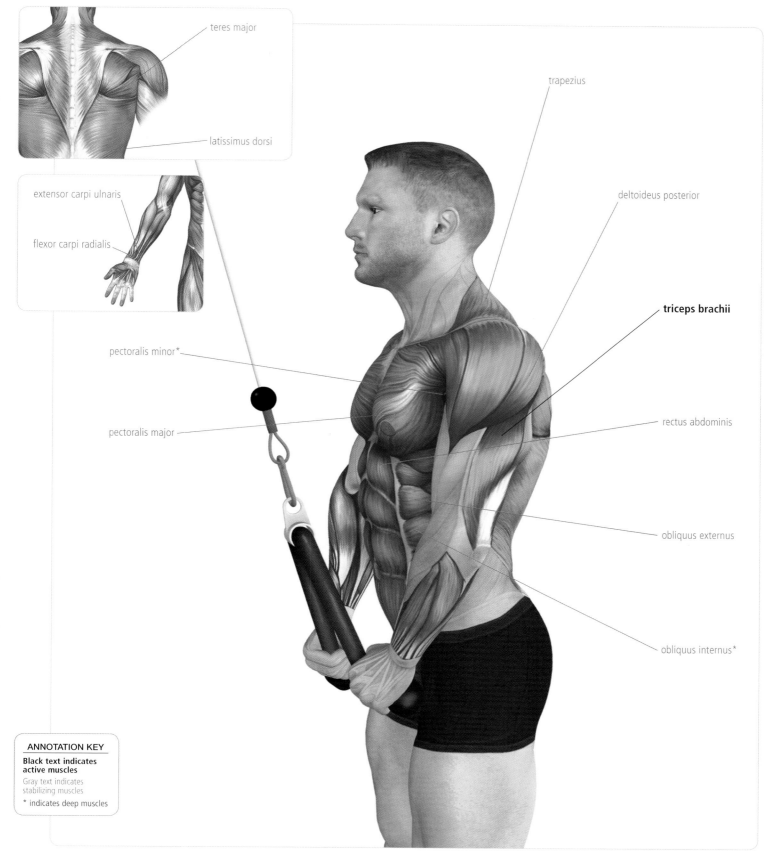

teres major

latissimus dorsi

extensor carpi ulnaris

flexor carpi radialis

trapezius

deltoideus posterior

triceps brachii

pectoralis minor*

pectoralis major

rectus abdominis

obliquus externus

obliquus internus*

ANNOTATION KEY

**Black text indicates
active muscles**

Gray text indicates
stabilizing muscles

* indicates deep muscles

ROPE HAMMER CURL

a

1. Attach a rope attachment to the lowest setting of a cable machine.

2. Stand about one foot from the cable upright with your feet parallel and shoulder width, your knees slightly bent, and your pelvis tucked in slightly. Grasp the rope with both hands in hammer-grip position, keeping your elbows tight to the sides of your body.

MUSCLES USED

- biceps brachii
- brachialis
- brachioradialis
- deltoideus anterior
- trapezius
- levator scapulae

b

3. Raise the cable weight toward your upper chest, keeping your upper arms stationary.

4. Slowly lower the weight back to the starting position, and repeat.

TARGET
- Biceps

LOOK FOR
- Your upper arms to remain stationary throughout the exercise.
- Your wrists to stay in line with your forearms.

AVOID
- Tensing your neck or jaw during exercise.

TRAINER'S TIPS
- For proper position, keep your neck back and your chin slightly up.
- Take a slight pause at the top range of the movement before lowering the cable weight back to the starting position.

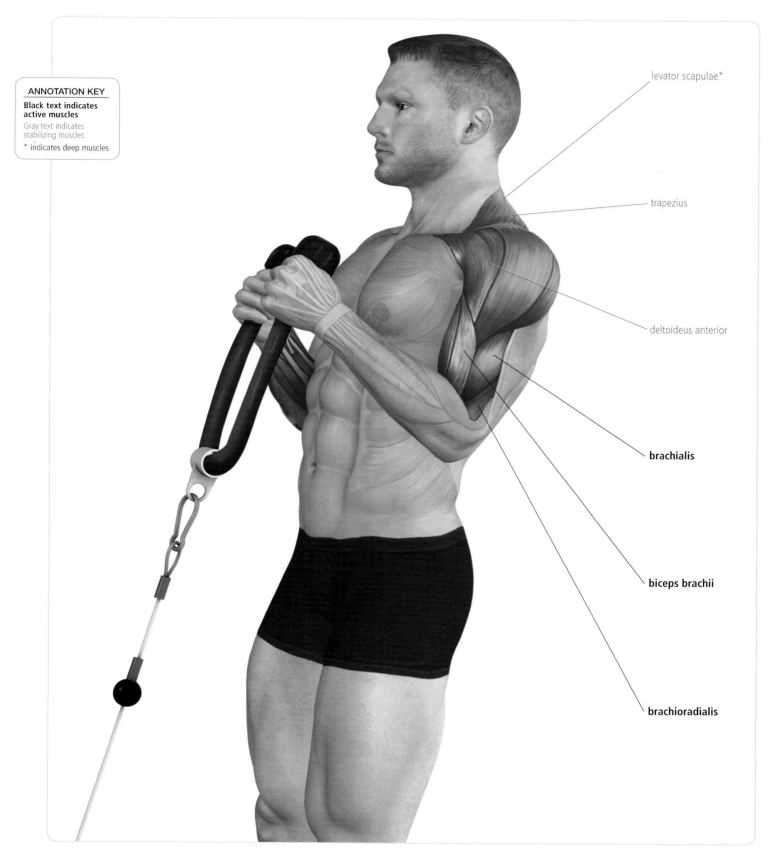

ANNOTATION KEY
Black text indicates active muscles
Gray text indicates stabilizing muscles
* indicates deep muscles

levator scapulae*

trapezius

deltoideus anterior

brachialis

biceps brachii

brachioradialis

RECLINING CABLE CURL

1 Attach an E-Z curl bar or straight bar attachment to the lowest setting of a cable machine pulley. Place an exercise mat on the floor in front of the cable machine.

2 With your arms shoulder width apart, grip the bar using both hands with palms facing upward.

3 Lie with your back flat on the mat, with your soles of your feet pressed against the frame of the pulley machine. This is your starting position, with your legs straight, arms extended and slightly bent, and elbows close to your body.

a

TARGET
• Biceps

LOOK FOR
• Your biceps to be fully contracted at the top range of the movement.

AVOID
• Bringing your head or lower back off the mat while executing the exercise.

MUSCLES USED

• biceps brachii	• trapezius
• brachioradialis	• levator scapulae
• brachialis	• flexor carpi radialis
• deltoideus anterior	• flexor carpi ulnaris

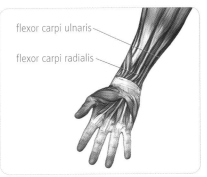

flexor carpi ulnaris

flexor carpi radialis

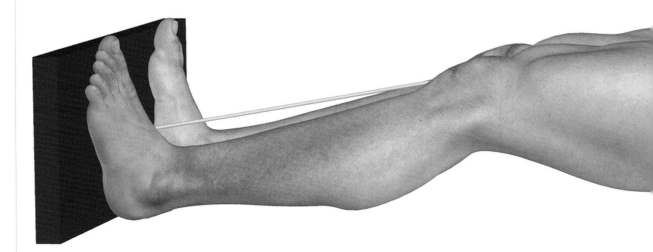

5 Keeping the arms stationary, curl the bar up toward your chest.

6 After a slight pause at the top, slowly lower the bar to the starting position, and repeat.

ⓑ

TRAINER'S TIPS
- Focus on squeezing your biceps during this exercise.
- Exhale as you bring the bar toward your chest, and inhale as you lower it back to the starting position.

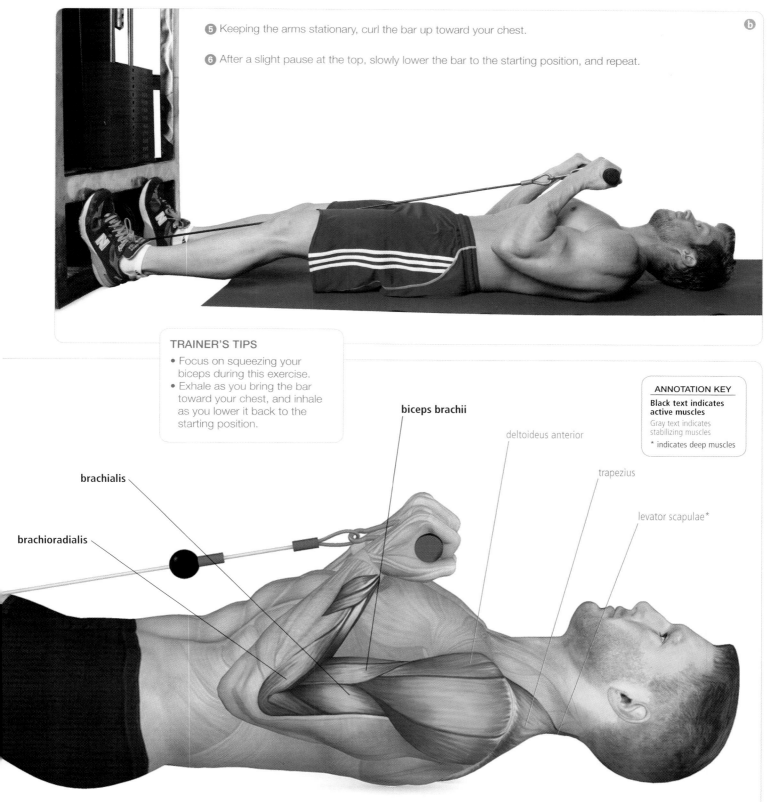

biceps brachii

deltoideus anterior

trapezius

levator scapulae*

brachialis

brachioradialis

ANNOTATION KEY
Black text indicates active muscles
Gray text indicates stabilizing muscles
* indicates deep muscles

BARBELL CURL

1 Stand with your feet parallel and shoulder width apart, your knees slightly bent and your pelvis slightly tucked in. Grip a barbell with your hands shoulder width apart and palms facing upward, elbows close to the sides of your body.

a

MUSCLES USED

- biceps brachii
- brachioradialis
- brachialis
- deltoideus anterior
- trapezius
- levator scapulae
- flexor carpi radialis
- flexor carpi ulnaris

2 Keeping your upper arms stationary, curl the barbell up toward your upper chest.

3 Take a slight pause at the top of the range of movement, and then slowly lower the barbell back to the starting position, and repeat.

TARGET
- Biceps

LOOK FOR
- Your upper arms to remain stationary throughout exercise.

AVOID
- Bending at the wrists—keep your wrists aligned with your forearms.
- Raising your shoulders.

TRAINER'S TIPS
- Exhale as you lift the barbell up, and inhale as you lower it back to the starting position.
- If you find that you are swaying your back or leaning too far backward, lean your back against a secure wall and place your feet about 2 feet in front of you. This will help you isolate the biceps.

b

CLOSE-GRIP MODIFICATION
Similar difficulty: Grip the barbell with palms facing upward with about 12 inches of space between your hands.

flexor carpi ulnaris

flexor carpi radialis

brachialis

brachioradialis

levator scapulae*

trapezius

deltoideus anterior

biceps brachii

WIDE-GRIP MODIFICATION
Similar difficulty: Grip the barbell with palms facing upward and your hands positioned outside of shoulder width, as far as is comfortable.

ANNOTATION KEY
Black text indicates active muscles
Gray text indicates stabilizing muscles
* indicates deep muscles

ALTERNATING HAMMER CURL

1 Stand with your feet parallel and shoulder width apart, your knees slightly bent and your pelvis slightly tucked in.

2 Grasp a dumbbell in each hand, using hammer-grip position. Your elbows should be close to the torso.

a

4 Slowly lower the weight back to the starting position, and repeat with the left dumbbell. Repeat, alternating sides.

c

TARGET
• Biceps

LOOK FOR
• Your biceps to be fully contracted at the top range of the movement.

AVOID
• Using momentum to lift the weight—keep your torso upright and concentrate on isolating and engaging the biceps.
• Bending at the wrists—keep your wrists aligned with your forearms.

b

3 With your upper arms remaining stationary, curl the right dumbbell toward your upper chest.

TRAINER'S TIPS
• Exhale as you lift the dumbbell up, and inhale as you lower it back to the starting position.
• If you find that you are swaying your back or leaning too far backward, lean your back against a secure wall and place your feet about two feet in front of you. This will help you isolate the biceps.

MUSCLES USED

• biceps brachii	• trapezius
• brachioradialis	• levator scapulae
• brachialis	• flexor carpi radialis
• deltoideus anterior	• flexor carpi ulnaris

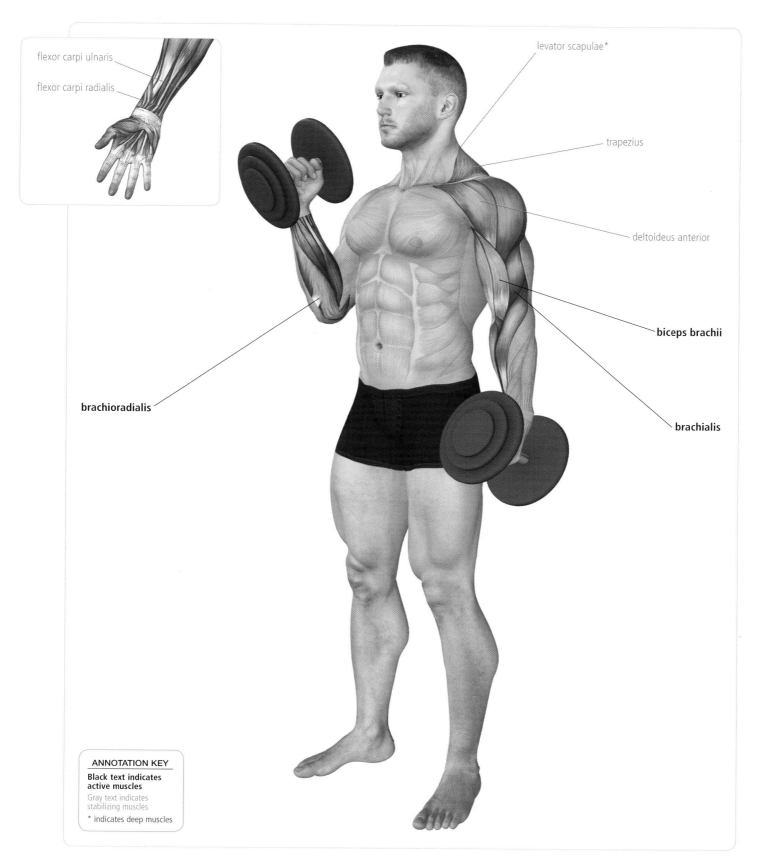

flexor carpi ulnaris

flexor carpi radialis

levator scapulae*

trapezius

deltoideus anterior

biceps brachii

brachialis

brachioradialis

DUMBBELL CURL

1 Straddle an incline bench facing toward the back. Grasp a dumbbell in each hand, palms facing each other.

2 Lean forward and carefully place dumbbells on the bench.

3 Roll the dumbbells off the bench, carefully lowering your body onto the bench. This is your starting position.

b

a

4 With your palms facing upward and your upper arms remaining stationary, curl the weights up toward the bench.

5 Slowly lower the weights back to the starting position, and repeat.

TARGET
• Biceps

LOOK FOR
• Your pelvis to tuck forward—this will support your lower back and help you avoid "sway back" while you execute this exercise.
• Your head to align with your spine and your chin to remain up.

AVOID
• Sliding down the incline bench—get a good solid grip on the floor with your feet.

MUSCLES USED

• biceps brachii	• trapezius
• brachioradialis	• levator scapulae
• brachialis	• flexor carpi radialis
• deltoideus anterior	• flexor carpi ulnaris

TRAINER'S TIPS
• Take a slight pause at the top of the movement.
• Exhale as you curl the dumbbells toward your chest, and inhale as you slowly lower them back to the starting position.

c

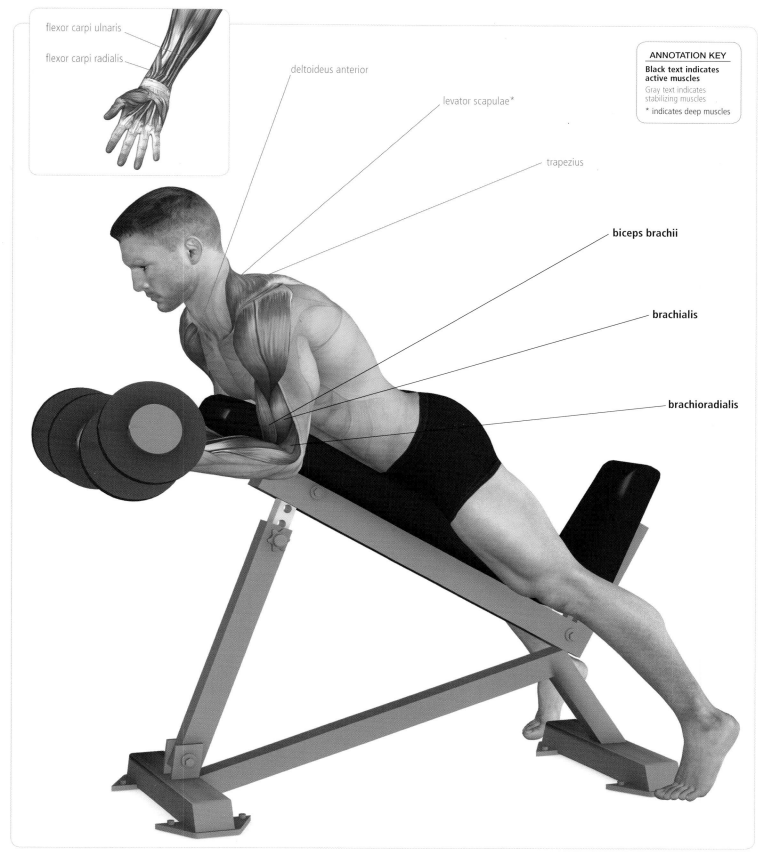

flexor carpi ulnaris

flexor carpi radialis

deltoideus anterior

levator scapulae*

trapezius

biceps brachii

brachialis

brachioradialis

PLATE CURL

1 Stand with your feet parallel and shoulder width apart, your knees slightly bent, and your pelvis slightly tucked in.

2 Grip a plate with both hands in hammer-grip position. Keep your upper arms stationary, your elbows close to your torso, and your shoulders down and back away from your ears.

a

MUSCLES USED

- biceps brachii
- brachioradialis
- brachialis
- deltoideus anterior
- trapezius
- levator scapulae
- flexor carpi radialis
- flexor carpi ulnaris

3 Curl the plate up toward your chest, pausing slightly at the top of the range of movement.

4 Slowly lower the plate back to the starting position, and repeat.

b

TARGET
- Biceps

LOOK FOR
- A smooth, even arc as you curl the plate upward and lower it back to the starting position.

AVOID
- Using your back to lift the plate—concentrate on isolating the biceps during this exercise.

TRAINER'S TIPS
- Use a 10-pound, 25-pound, or 45-pound plate, depending on your strength and number of repetitions.

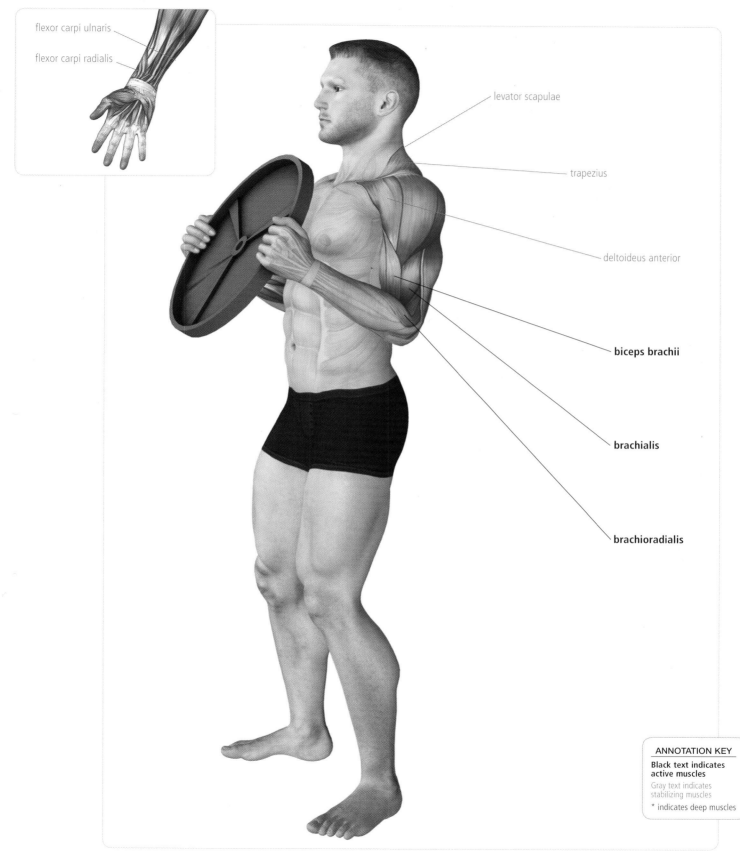

flexor carpi ulnaris

flexor carpi radialis

levator scapulae

trapezius

deltoideus anterior

biceps brachii

brachialis

brachioradialis

ANNOTATION KEY

Black text indicates active muscles

Gray text indicates stabilizing muscles

* indicates deep muscles

SINGLE-ARM CONCENTRATION CURL

1 Sit facing forward on a flat bench with your legs spread generously outside of shoulder width. Hold a dumbbell in front of you between your legs.

2 Rest the back of your right upper arm on the top of your inner right thigh.

a

MUSCLES USED

- biceps brachii
- brachioradialis
- brachialis
- trapezius
- levator scapulae
- flexor carpi radialis
- flexor carpi ulnaris
- obliquus externus
- obliquus internus
- erector spinae

TRAINER'S TIPS

- Exhale while curling the dumbbell upward, and inhale as you lower it back to the starting position.
- Take a slight pause at the top range of the movement, concentrating on the contraction of the biceps.
- To ensure a good grip, contraction of the biceps, and good wrist alignment, squeeze your pinky finger. Your pinky should be higher than your thumb.

3 With your right palm facing upward and your upper arm stationary, curl the dumbbell forward and up toward your face, stopping at shoulder height.

4 Slowly lower the weight back to the starting position and repeat.

5 Switch sides, and repeat all steps with your left hand holding the dumbbell.

TARGETS
- Brachialis

LOOK FOR
- The dumbbell to be a few inches from the floor when the arm is extended down in the starting position.

AVOID
- Swinging motions during this exercise.
- Bending your wrist—keep your wrist aligned with your forearm.
- Rolling your shoulder inward.

b

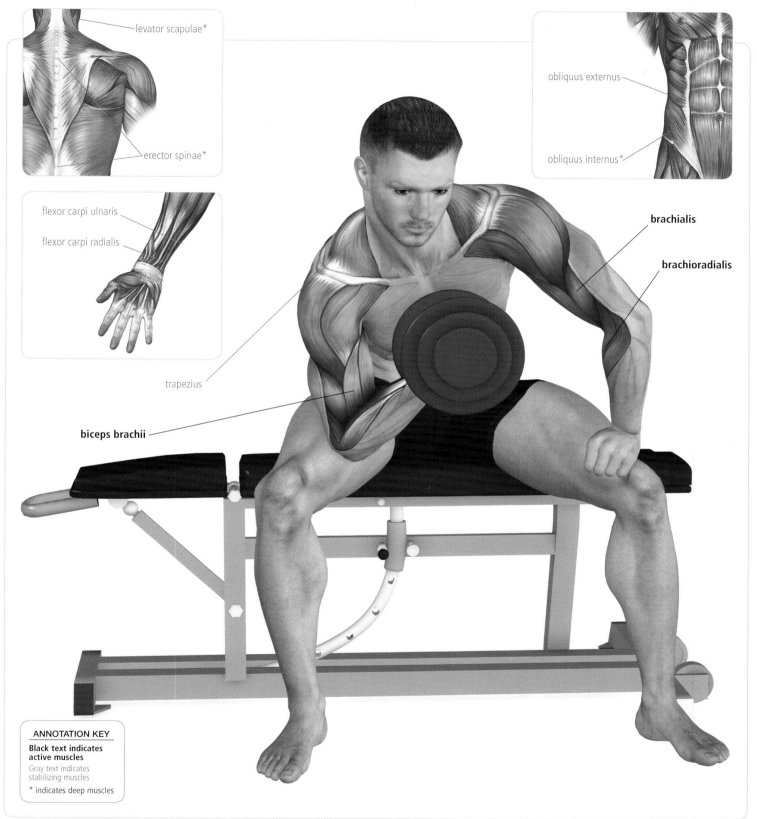

levator scapulae*

erector spinae*

obliquus externus

obliquus internus*

brachialis

brachioradialis

flexor carpi ulnaris

flexor carpi radialis

trapezius

biceps brachii

ANNOTATION KEY

Black text indicates active muscles

Gray text indicates stabilizing muscles

* indicates deep muscles

LEGS

The major leg muscles can be divided into three groups: the quadriceps femoris (the strongest muscle of the body), the hamstrings, and the calves.

The quadriceps femoris actually consists of four front thigh muscles—the vastus lateralis, vastus medialis, vastus intermedius, and rectus femoris. The hamstrings are muscles of the back thigh—the semitendinosus, semimembranosus, and biceps femoris. The semitendinosus and semimembranosus extend the hip when the trunk is fixed, and they flex the knee and rotate the lower leg inward when the knee is bent. The biceps femoris extends the hip when we take a step to walk, and it also flexes the knee and rotates the lower leg outward when the knee is bent. Both of these groups allow us to walk, run, jump, and squat.

The major calf muscles are the gastrocnemius and the soleus. The gastrocnemius has two heads that when fully developed form a distinctive diamond shape. Both the gastrocnemius and soleus work to raise the heel.

SMITH MACHINE SQUAT

1 Set the barbell of a Smith machine just below the height of your shoulders, and load up the barbell with the appropriate weight. Standing with feet parallel and your feet shoulder width apart, place the bar across the back of your shoulders, slightly below your neck.

2 Hold the bar with your hands spread wide apart from each other and palms facing forward. Unlock the rack by pushing up with your legs and straightening your torso.

a

MUSCLES USED

- rectus femoris
- vastus lateralis
- vastus intermedius
- adductor magnus
- gluteus maximus
- soleus
- biceps femoris
- semitendinosus
- semimembranosus
- erector spinae
- gastrocnemius
- obliquus externus
- obliquus internus

TARGETS
- Quadriceps

LOOK FOR
- Your head to remain up at all times to ensure proper posture.

AVOID
- Allowing your knees to go past your toes as you bend—this will exert great stress on the knee and may cause injury.
- Rolling your shoulders or upper back forward during this exercise.

3 Step both legs forward, until you are just slightly leaning backward. Your feet should be shoulder width apart and your toes slightly pointing out.

4 Slowly lower the weight by bending your knees until the angle between your thighs and calves becomes slightly less than 90 degrees.

5 Push your heels into the floor as you straighten your legs back to the starting position, and repeat.

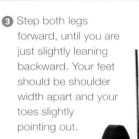

b

TRAINER'S TIPS

- Inhale as you lower the weight during your squat, and exhale as you push through your heels to raise back to the starting position.
- If you suffer from back issues, first try with a leg press machine.
- This exercise is the same as a Barbell Squat, but uses a Smith machine to support a greater weight, allowing you to lift heavier.
- A neck pad may be available at your gym for added comfort.
- When you are finished with this exercise, carefully step both feet back underneath you, bend your knees slightly, and rack the weight back onto the Smith machine just below shoulder height.
- This is an advanced exercise; perform it with care and perfect form.

erector spinae*

gluteus maximus

adductor magnus

biceps femoris

semitendinosus

semimembranosus

gastrocnemius

obliquus externus

obliquus internus*

vastus intermedius*

rectus femoris

vastus lateralis

soleus

ANNOTATION KEY

Black text indicates active muscles

Gray text indicates stabilizing muscles

* indicates deep muscles

SMITH MACHINE SINGLE-LEG LUNGE

1 Set the barbell of a Smith machine just below the height of your shoulders, and load up the barbell with the appropriate weight. Standing with feet parallel and your feet shoulder width apart, place the bar across the back of your shoulders, slightly below your neck.

2 Hold the bar with both hands with palms facing forward. Unlock the rack by pushing up with your legs and straightening your torso.

3 Step forward with your right leg about one-half to one foot in front of you, and then extend your left leg back about one-half to one foot behind you. This is your starting lunge position.

a

b

TARGETS
• Quadriceps

LOOK FOR
• Your head to remain up at all times to ensure proper posture.

AVOID
• Rolling your knee in as you come up from your lunge. Keep your knee over your toes as you push your heel into the floor to raise back to the starting position.
• Rolling your shoulders or upper back forward during this exercise.

4 Slowly lower the weight by bending your right knee until the angle between your thigh and calf becomes slightly less than 90 degrees.

5 Push down into your right heel as you come up to the starting lunge position, and repeat for the desired number of reps.

6 Step the left leg back underneath yourself, then your right leg back underneath you. To switch legs, bring your left leg forward and the right leg back.

TRAINER'S TIPS
• Inhale as you lower the weight during your lunge, and exhale as you raise back to your starting lunge position.
• When you are finished with this exercise, bring both legs back underneath you and bending slightly from the knees, rack the weight back onto the Smith machine just below shoulder height.
• A neck pad might be available at your gym for added comfort.
• If you suffer from back issues or knee issues, you should first attempt dumbbell lunges.
• This is an advanced exercise; perform it with care and perfect form.

c

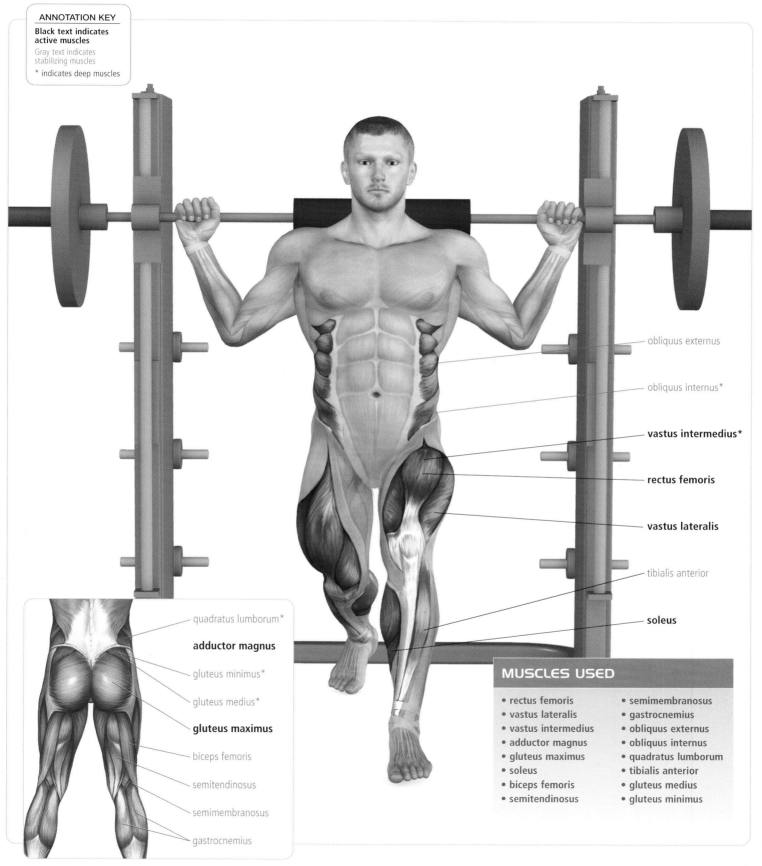

ANNOTATION KEY

Black text indicates active muscles

Gray text indicates stabilizing muscles

* indicates deep muscles

obliquus externus

obliquus internus*

vastus intermedius*

rectus femoris

vastus lateralis

tibialis anterior

soleus

quadratus lumborum*

adductor magnus

gluteus minimus*

gluteus medius*

gluteus maximus

biceps femoris

semitendinosus

semimembranosus

gastrocnemius

MUSCLES USED

- rectus femoris
- vastus lateralis
- vastus intermedius
- adductor magnus
- gluteus maximus
- soleus
- biceps femoris
- semitendinosus
- semimembranosus
- gastrocnemius
- obliquus externus
- obliquus internus
- quadratus lumborum
- tibialis anterior
- gluteus medius
- gluteus minimus

FLAT BENCH DUMBBELL SQUAT

1 Stand about one to two feet in front of a flat bench with your feet parallel and placed generously outside of shoulder width. Grasp a dumbbell in each hand using the hammer-grip positioned, palms facing each other.

a

MUSCLES USED

- rectus femoris
- vastus lateralis
- vastus intermedius
- vastus lateralis
- soleus
- biceps femoris
- semitendinosus
- semimembranosus
- erector spinae
- trapezius
- levator scapulae
- gastrocnemius
- obliquus externus
- obliquus internus
- rectus abdominis

TRAINER'S TIPS

- Focus on engaging your glutes and thigh muscles.
- To engage the biceps, keep a slight bend in your elbows.
- Exhale as you squat down, and inhale as you raise back to the starting position.
- Think of this exercise as an etiquette class in which you must balance a book on the top of your head. This will ensure that you attain proper upper-body posture while executing your squat.

TARGETS
- Quadriceps
- Gluteal muscles

LOOK FOR
- Your torso to remain upright for the duration of this exercise.
- Your head to remain aligned with your spine and your chin to remain slightly up for proper form.

AVOID
- Rolling your shoulders and upper back forward.

2 Squat down toward the bench, lightly touching the bench with your glutes.

3 Slowly raise back to the standing start position, and repeat.

b

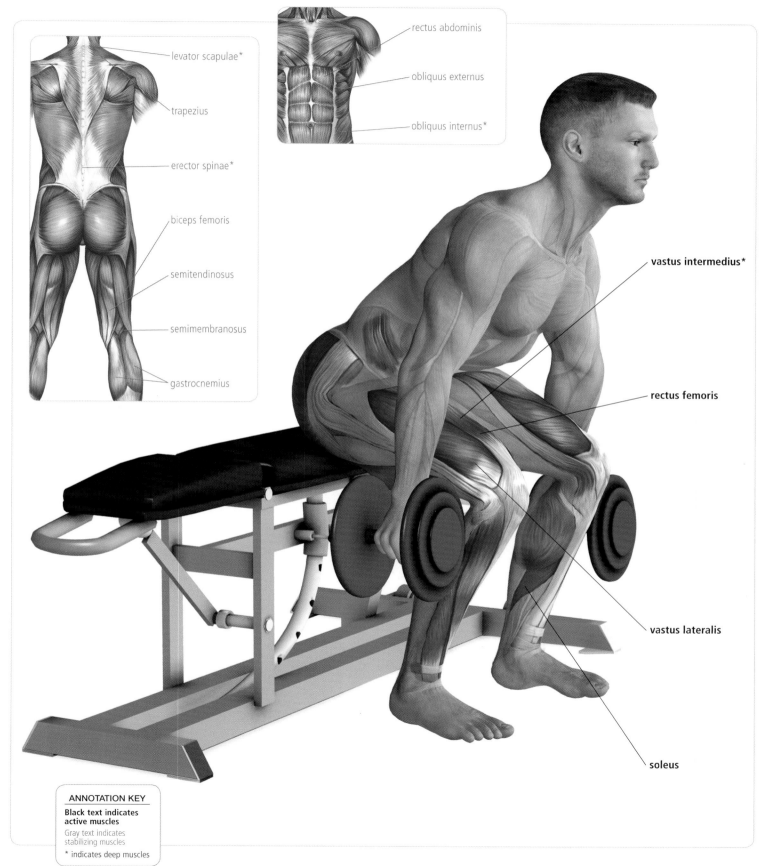

rectus abdominis

obliquus externus

obliquus internus*

levator scapulae*

trapezius

erector spinae*

biceps femoris

semitendinosus

semimembranosus

gastrocnemius

vastus intermedius*

rectus femoris

vastus lateralis

soleus

ANNOTATION KEY

Black text indicates active muscles

Gray text indicates stabilizing muscles

* indicates deep muscles

DUMBBELL WALKING LUNGE

TRAINER'S TIPS

• Inhale as you step forward and exhale as you raise back to the starting position.
• Think of this exercise as an etiquette class in which you must balance a book on the top of your head. This will ensure that you attain proper upper-body posture while executing your lunge.
• This is a very advanced exercise; perform it with care and correct form.

1 Stand with your feet parallel and slightly narrower than shoulder width apart, holding a dumbbell in each hand in hammer-grip position, palms facing each other. Keep your arms close to the sides of your body.

2 Step forward with your left leg until your left foot is approximately two feet from your right foot, keeping your torso upright as you lower your upper body.

3 Concentrating on using your left heel, push up and forward, returning to your starting stance position.

4 Repeat steps 2 and 3 starting with right leg.

TARGETS
• Quadriceps
• Gluteal muscles

LOOK FOR
• Your front shin to remain perpendicular to the ground.
• Your torso to remain upright for the duration of this exercise.

AVOID
• Allowing the stepping knee to go forward beyond your toes as you lower down—this could stress the knee joint and cause possible injury.

MUSCLES USED

- rectus femoris
- vastus lateralis
- vastus intermedius
- adductor magnus
- gluteus maximus
- soleus
- biceps femoris
- semitendinosus
- semimembranosus
- erector spinae
- gastrocnemius
- obliquus externus
- obliquus internus
- quadratus lumborum
- tibialis anterior
- gluteus minimus
- gluteus medius

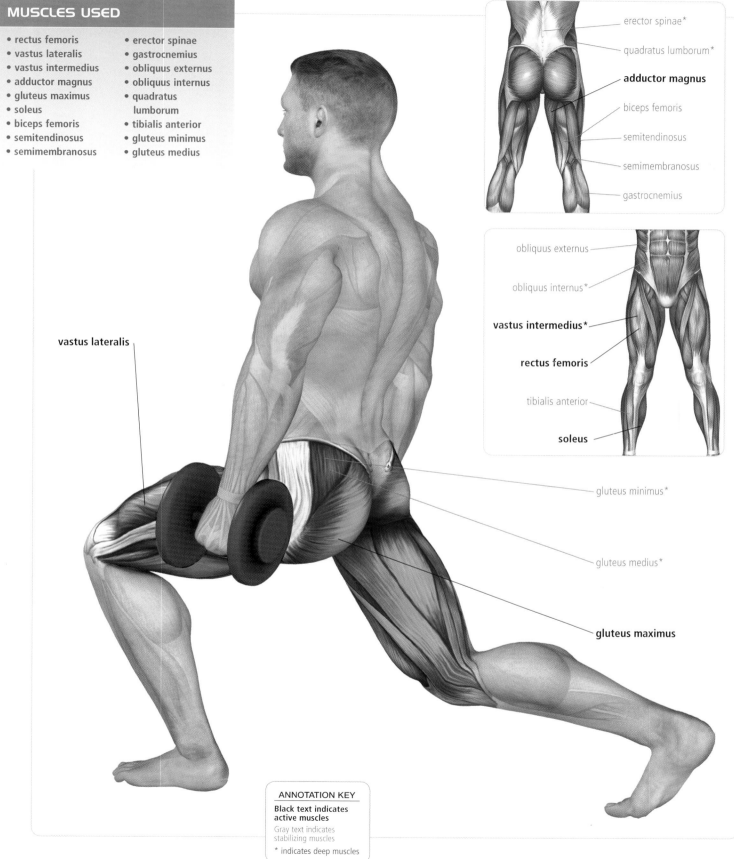

erector spinae*

quadratus lumborum*

adductor magnus

biceps femoris

semitendinosus

semimembranosus

gastrocnemius

obliquus externus

obliquus internus*

vastus intermedius*

rectus femoris

tibialis anterior

soleus

vastus lateralis

gluteus minimus*

gluteus medius*

gluteus maximus

ANNOTATION KEY

Black text indicates active muscles

Gray text indicates stabilizing muscles

* indicates deep muscles

FLAT BENCH STEP-UP

a

1 Stand about one foot behind a flat bench with your feet parallel and close together. Grip a dumbbell in each hand.

b

2 Step up with the right leg up onto the bench.

TRAINER'S TIPS
- Focus on engaging your glutes and thigh muscles.
- Exhale as you step up onto the bench, and inhale as you step down.

c

3 Bring the left leg up beside the right leg.

d

4 Step down with the right leg, and then step down with the left leg to return to the starting position.

5 Continue leading with the right foot up and off the bench for as many reps as your choose. Then switch legs, and lead up with the left leg, and lead down with the left leg.

TARGETS
- Quadriceps
- Gluteal muscles
- Hamstrings

LOOK FOR
- Your torso to remain upright for the duration of this exercise.
- A slow, even, steady pace with both the step-ups and the step-downs.

AVOID
- Rolling your shoulders and upper back forward while stepping up onto the bench.

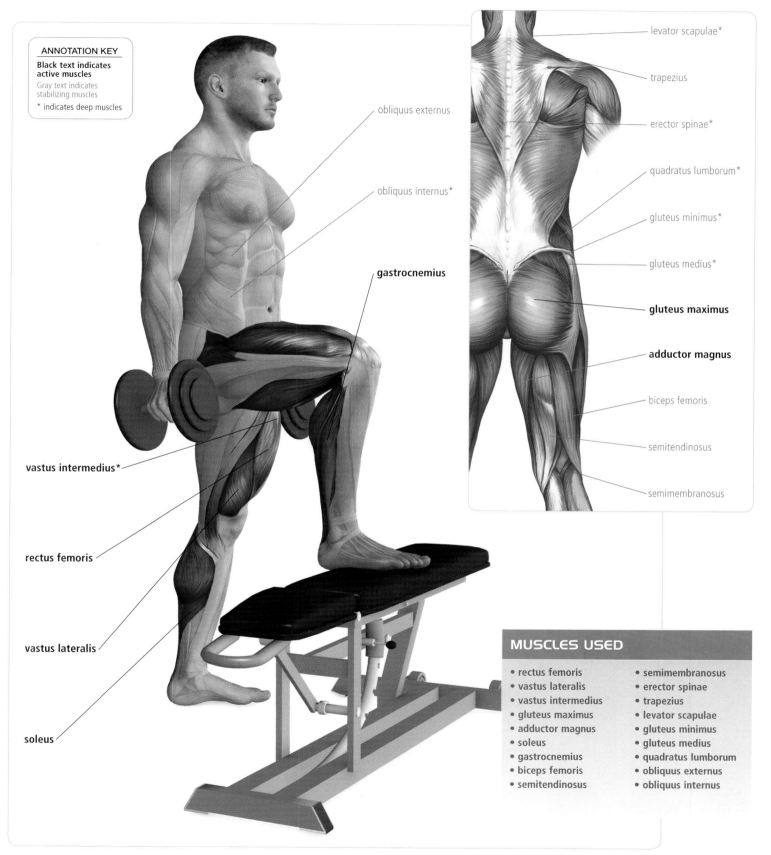

ANNOTATION KEY

Black text indicates active muscles

Gray text indicates stabilizing muscles

* indicates deep muscles

obliquus externus

obliquus internus*

gastrocnemius

vastus intermedius*

rectus femoris

vastus lateralis

soleus

levator scapulae*

trapezius

erector spinae*

quadratus lumborum*

gluteus minimus*

gluteus medius*

gluteus maximus

adductor magnus

biceps femoris

semitendinosus

semimembranosus

MUSCLES USED

- rectus femoris
- vastus lateralis
- vastus intermedius
- gluteus maximus
- adductor magnus
- soleus
- gastrocnemius
- biceps femoris
- semitendinosus
- semimembranosus
- erector spinae
- trapezius
- levator scapulae
- gluteus minimus
- gluteus medius
- quadratus lumborum
- obliquus externus
- obliquus internus

BARBELL SQUAT

① To begin, set a bar on a squat rack just below your shoulder level. Load the bar with proper weights, step under the bar, and place the back of your shoulders slightly below the neck under the bar. Hold onto the bar using both arms out across the bar.

② Standing with your knees bent slightly, lift the barbell off the rack. Carefully take a step back.

③ Position your feet so that they are parallel and shoulder width apart, maintaining a slight bend of the knees. This is your starting position.

a

MUSCLES USED

- rectus femoris
- vastus lateralis
- vastus intermedius
- adductor magnus
- gluteus maximus
- soleus
- biceps femoris
- semitendinosus
- semimembranosus
- erector spinae
- gastrocnemius
- obliquus externus
- obliquus internus
- supraspinatus
- pectoralis major
- trapezius
- levator scapulae
- serratus anterior
- rectus abdominis

TARGETS
- Quadriceps
- Hamstrings

LOOK FOR
- Your head to remain aligned with your spine and your chin to remain slightly up for proper form.
- The bar to remain straight and balanced during exercise to ensure safety.
- Your spine to remain stable and from head to hips.

AVOID
- Rolling the shoulders or upper back forward during this exercise.

④ Keeping your back straight, slowly bend your knees until your thighs and calves form an angle of slightly less than 90 degrees.

⑤ Pushing down into your heels, begin to raise the bar by straightening your legs back to the upright starting position. Repeat.

TRAINER'S TIPS
- Inhale as you squat, and exhale as you lift and return to the starting position.
- This exercise is safest while being performed within a squat rack. When you are finished with this exercise, carefully step forward, aligning the bar back into the rack. Secure the weight onto the rack by bending your knees slightly.
- Perform this exercise with extreme caution. If you have back issues, attempt squats with proper dumbbell weight first.

b

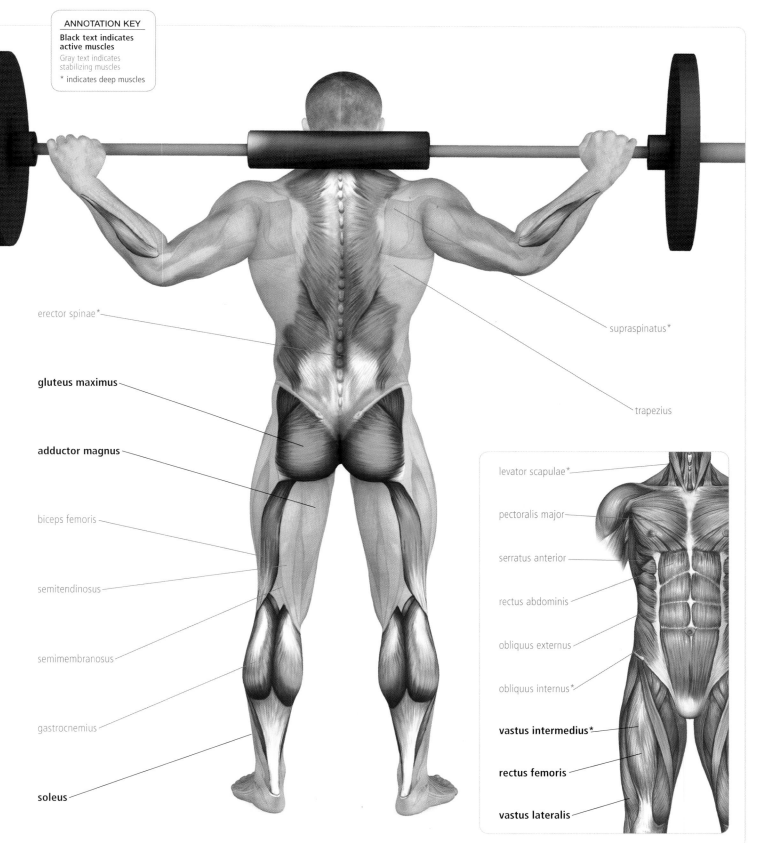

ANNOTATION KEY

Black text indicates active muscles

Gray text indicates stabilizing muscles

* indicates deep muscles

erector spinae*

gluteus maximus

adductor magnus

biceps femoris

semitendinosus

semimembranosus

gastrocnemius

soleus

supraspinatus*

trapezius

levator scapulae*

pectoralis major

serratus anterior

rectus abdominis

obliquus externus

obliquus internus*

vastus intermedius*

rectus femoris

vastus lateralis

STIFF-LEGGED BARBELL DEADLIFT

1 Stand with your feet parallel and shoulder width apart, with a barbell in front of you on the ground. Keeping your back as straight as possible, bend forward and grasp the bar with an overhand grip, palms facing downward.

a

2 With your knees straight or only very slightly bent and shins kept vertical, hips back, and back straight, use your hips to lift the bar.

b

3 Continue lifting until you are in a standing position.

4 Lower the weight back to the starting position, making sure to keep the barbell close to the front on the body.

TARGETS
- Hamstrings
- Gluteal muscles
- Lower back

LOOK FOR
- A slightly faster movement than with other exercises.
- Steady but controlled movement—safety and proper form is necessary.

AVOID
- Rounding the back forward when you perform this exercise.
- Using momentum to raise and lower the barbell.

TRAINER'S TIPS
- Inhale as you lower the barbell, and exhale as you lift back to the starting position.
- If you find holding onto the bar difficult, wrist straps will secure the bar and allow you to lift a heavier weight.
- This exercise can also be performed using dumbbells in each hand.
- Avoid this exercise if you suffer from lower-back problems.

c

MUSCLES USED

- biceps femoris
- semitendinosus
- semimembranosus
- gluteus maximus
- erector spinae
- trapezius
- rhomboideus
- latissimus dorsi
- levator scapulae
- rectus abdominis
- obliquus externus
- obliquus internus

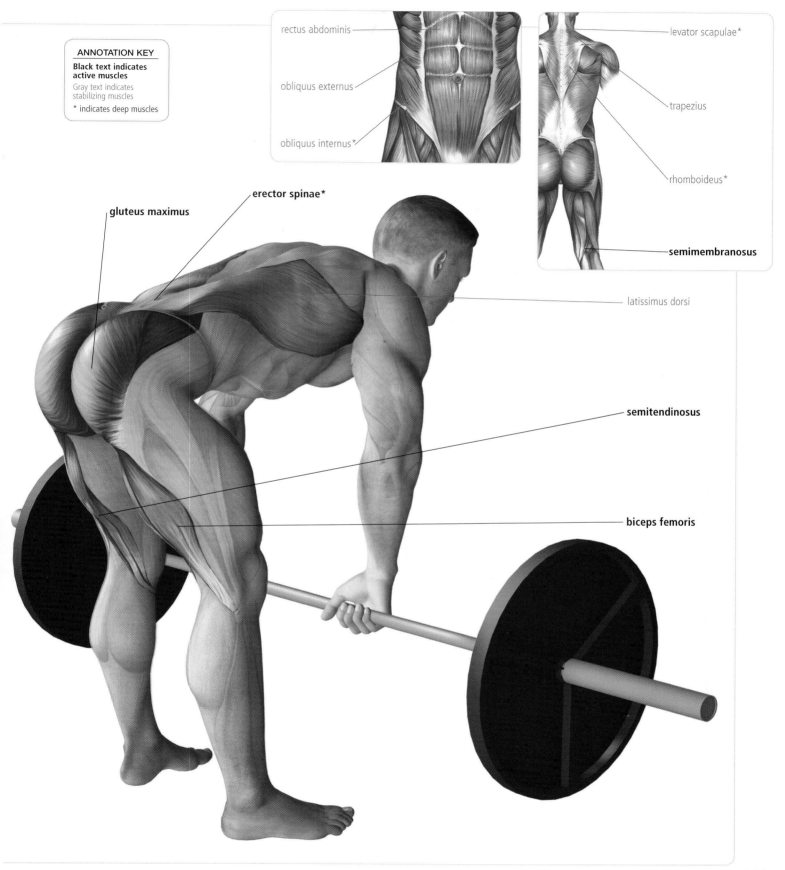

ANNOTATION KEY

**Black text indicates
active muscles**

Gray text indicates
stabilizing muscles

* indicates deep muscles

rectus abdominis

obliquus externus

obliquus internus*

levator scapulae*

trapezius

rhomboideus*

semimembranosus

gluteus maximus

erector spinae*

latissimus dorsi

semitendinosus

biceps femoris

PLIÉ SQUAT

LEGS

① Stand with your feet turned out slightly and wider than shoulder width apart and your knees bent slightly, grasping a single dumbbell at the base with both hands.

a

b

② Slowly bend the knees and lower your legs until your thighs are parallel to the floor.

③ Slowly raise back up to the starting position, and repeat.

TARGETS
• Hamstrings
• Gluteal muscles

LOOK FOR
• Your torso to remain upright for the duration of this exercise.

AVOID
• Moving your arms during this exercise.
• Rolling the shoulders or upper back forward during this exercise.

TRAINER'S TIPS

• Inhale as you lower into the squat, and exhale as you raise back to the starting position.
• Concentrate on pressing down with your heels as you raise your body back to the starting position.
• Make sure your knees are in line with the direction of your toes. Turning out from the thighs will help ensure this proper positioning.
• This exercise can really work those inner and outer thigh muscles that do not receive the attention in a traditional squat.

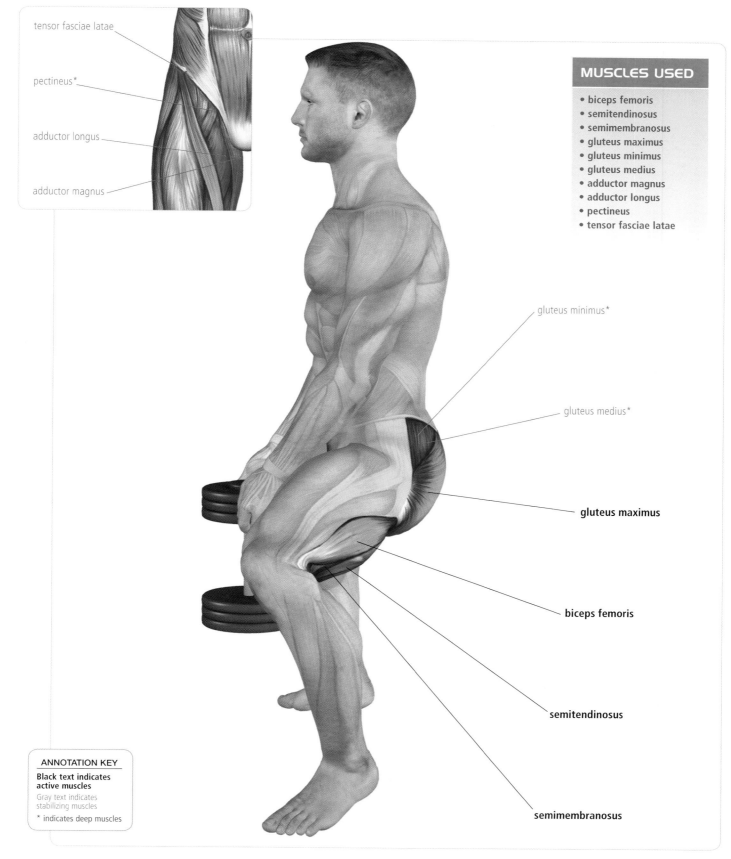

tensor fasciae latae

pectineus*

adductor longus

adductor magnus

MUSCLES USED

- biceps femoris
- semitendinosus
- semimembranosus
- gluteus maximus
- gluteus minimus
- gluteus medius
- adductor magnus
- adductor longus
- pectineus
- tensor fasciae latae

gluteus minimus*

gluteus medius*

gluteus maximus

biceps femoris

semitendinosus

semimembranosus

ANNOTATION KEY

Black text indicates active muscles

Gray text indicates stabilizing muscles

* indicates deep muscles

DUMBBELL SHIN RAISE

1 Sit on the front edge of a flat bench with a dumbbell on the floor in front of you. Clasp the dumbbell with your feet.

a

2 Shimmy back onto the bench so that only your feet hang off the bench. Keeping your legs straight and torso sitting up straight, point the feet slowly.

b

c

TARGETS
• Shins

LOOK FOR
• A full range of movement while pointing and flexing the feet.
• Your neck and jaw to remain relaxed throughout the exercise.

AVOID
• Bending the knees while performing this exercise.

3 Then, keeping your legs straight and torso sitting up straight, flex the feet slowly. Repeat.

TRAINER'S TIPS
• Inhale as you point your feet, and exhale as you flex them.
• When you are finished with this exercise, carefully lower the dumbbell to the floor.

ANNOTATION KEY

Black text indicates active muscles

Gray text indicates stabilizing muscles

* indicates deep muscles

• tibialis anterior

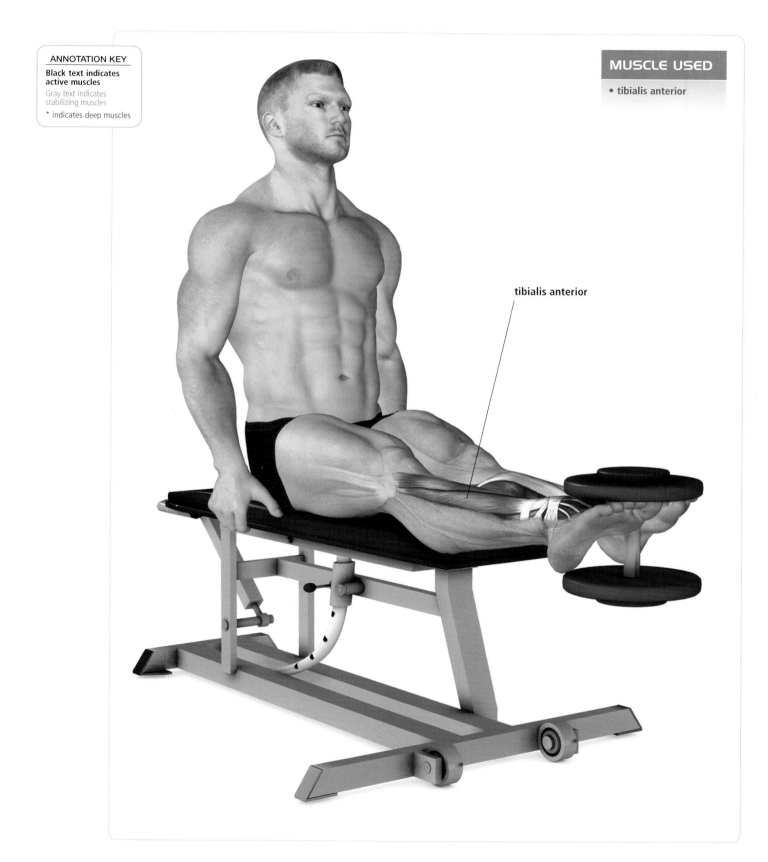

tibialis anterior

DUMBBELL CALF RAISE

1 Stand with your feet parallel and shoulder width apart, holding a dumbbell in each hand in hammer-grip position, palms facing each other. Keep your arms close to the sides of your body.

2 Slowly raise onto the balls of your feet, concentrating on contracting your calf muscles as you raise.

3 Slowly lower back to the starting position, and repeat.

a

b

TARGETS
• Calves

LOOK FOR
• A strong contraction of the calf muscles during this exercise.

AVOID
• Bending your knees while performing this exercise.
• "Sickling" your foot as you raise onto your toes—meaning that you roll onto the smaller toes. Keep the focus over the big toe for proper form.

TRAINER'S TIPS
• Exhale as you raise onto your toes, and inhale as you lower back to the starting position.
• As you become stronger, wearing wrist straps to grip heavier weight will be a good option.
• For a more advanced version of this exercise, place two plates on the floor and give yourself a lift by placing the balls of your feet on the plates. This gives the calf muscles a greater stretch.

MUSCLES USED

• gastrocnemius
• soleus
• trapezius
• levator scapulae
• gluteus minimus
• gluteus medius

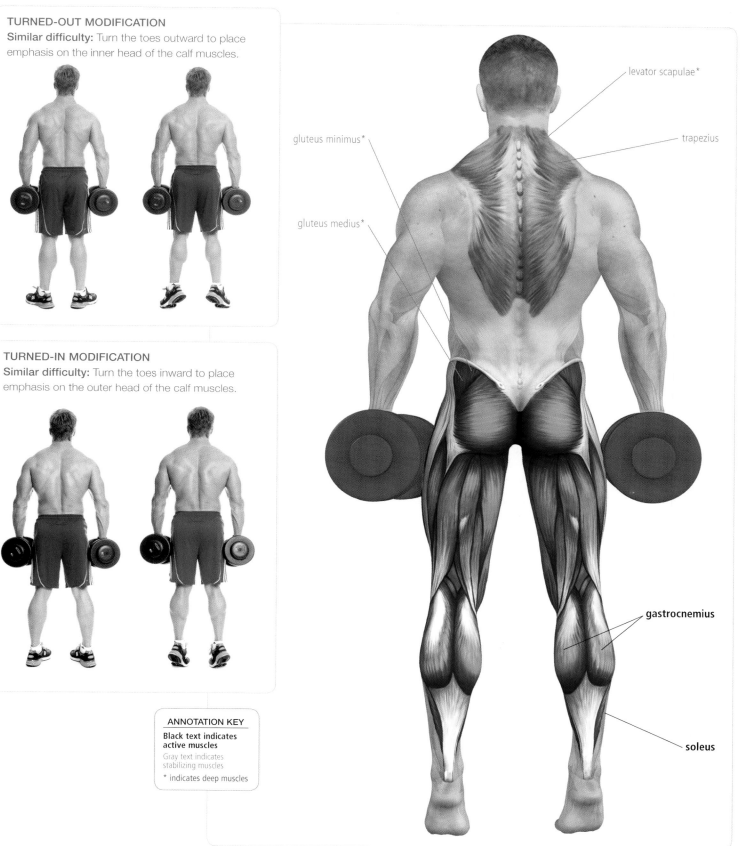

TURNED-OUT MODIFICATION

Similar difficulty: Turn the toes outward to place emphasis on the inner head of the calf muscles.

TURNED-IN MODIFICATION

Similar difficulty: Turn the toes inward to place emphasis on the outer head of the calf muscles.

ANNOTATION KEY

Black text indicates active muscles

Gray text indicates stabilizing muscles

* indicates deep muscles

levator scapulae*

trapezius

gluteus minimus*

gluteus medius*

gastrocnemius

soleus

GLOSSARY

GENERAL TERMS

abduction: Movement away from the body.

adduction: Movement toward the body.

alternating grip: One hand grasping with the palm facing toward the body and the other facing away.

anterior: Located in the front.

barbell: A basic piece of equipment consisting of a long bar, collar, sleeves, and associated plates made of steel or iron. An adjustable barbell allows the changing of plates and a fixed barbell features welded collars that keep the plates in place. A typical bar averages between 5 and 7 feet in length and usually weighs between 25 and 45 pounds.

bench press: An exercise movement that calls for a person to lie supine on a bench, lower a weight to chest level, and then push it back up until the arm is straight and the elbows are locked. Bench presses strengthen the pectorals, deltoids, and triceps.

bosu ball: A flexible dome-shaped platform with a flat bottom that resembles a stability ball chopped in half. Also called a "balance ball" or "balance trainer," it is used to perform strength, balance, coordination, and cardiovascular exercises.

cable machine: A piece of gym equipment that features long wire cords attached to weight stacks at one end and a hand grip at the other. A cable exercise keeps tension on the working muscle through the full range of motion.

cardiovascular exercise: Any exercise that increases the heart rate, making oxygen and nutrient-rich blood available to working muscles.

cardiovascular system: The circulatory system that distributes blood throughout the body, which includes the heart, lungs, arteries, veins, and capillaries.

collar: A small, round, iron or plastic clamp that anchors plates on a barbell or dumbbell.

concentric (contraction): Occurs when a muscle shortens in length and develops tension, e.g., the upward movement of a dumbbell in a biceps curl.

crunch: A common abdominal exercise that calls for curling the shoulders toward the pelvis while lying supine with hands behind head and knees bent.

curl: An exercise movement, usually targeting the biceps brachii, that calls for a weight to be moved through an arc, in a "curling" motion.

deadlift: An exercise movement that calls for lifting a weight, such as a barbell, off the ground from a stabilized bent-over position.

decline bench: A bench in which the user places his or her head at the low end and feet at the upper end. Often used to work the lower and outer pectorals

dumbbell: A basic piece of equipment that consists of a short bar on which plates are secured. A person can use a dumbbell in one had or in both hands during an exercise. Most gyms offer dumbbells with the weight plates welded on and poundage indicated on the plate.

dynamic: Continuously moving.

eccentric (contraction): The development of tension while a muscle is being lengthened, e.g., the downward movement of a dumbbell in a biceps curl.

extension: The act of straightening.

extensor muscle: A muscle serving to extend a body part away from the body.

EZ-curl bar: A short, S-shaped bar used for such exercises as skull crushers. The S-shape puts less stress on the wrists and forearms than a straight bar.

flexion: The bending of a joint.

flexor muscle: A muscle that decreases the angle between two bones, as bending the arm at the elbow or raising the thigh toward the stomach.

fly: An exercise movement in which the hand and arm move through an arc while the elbow is kept at a constant angle. Flyes work the muscles of the upper body.

free weights: Barbells and dumbbells.

hammer grip: Grasping dumbbells or exercise equipment with your palms facing inward toward each other.

iliotibial band (ITB): A thick band of fibrous tissue that runs down the outside of the leg, beginning at the hip and extending to the outer side of the tibia just below the knee joint. The band functions in coordination with several of the thigh muscles to provide stability to the outside of the knee joint.

incline bench: A bench in which the body is tilted back in respect to the vertical. Often used to work the upper-chest region.

lateral: Located on, or extending toward, the outside.

lordosis: Forward curvature of the spine and lumbar region.

medial: Located on, or extending toward, the middle.

muscle failure: Refusing to terminate a set of exercise repetitions until the muscle simply cannot contract for additional reps. Also called "training to failure."

neutral position (spine): A spinal position resembling an S shape, consisting of a lordosis in the lower back, when viewed in profile.

overhand grip: Grasping a barbell or dumbbell with the palms facing downward and the thumbs facing inward. Also called the "pronated grip," this is the most common weightlifting grip.

plate: A cast-iron weight placed on a barbell or dumbbell. Plates range in size from 1¼ pound to 100 pounds. The common plates found in muscle-building gyms weigh 5, 10, 25, 25, and 45 pounds.

posterior: Located behind.

press: An exercise movement that calls for moving a weight or other resistance away from the body.

pronated grip: See *overhand grip*.

reverse grip: Grasping a barbell or dumbbell with palms facing upward. Also called the "supinated grip."

scapula: The protrusion of bone on the mid to upper back, also known as the "shoulder blade."

Smith machine: A piece of weight-training equipment that consists of a barbell fixed within steel rails, allowing only vertical movement.

squat: An exercise movement that calls for moving the hips back and bending the knees and hips to lower the torso and an accompanying weight, and then returning to the upright position. A squat primarily targets the muscles of the thighs, hips and buttocks, and hamstrings.

stability ball: A flexile, inflatable PVC ball measuring approximately 14 to 34 inches in circumference that is used for weight training, physical therapy, balance training, and many other exercise regimens. It is also called a "Swiss ball," "fitness ball," "exercise ball," "gym ball," "physioball," and many other names.

supinated grip: See *reverse grip*.

training partner: An individual who works out with you, usually matching you set for set. A training partner acts as a motivator, and he or she also acts as a spotter when you lift heavy weights.

warm-up: Any form of light exercise of short duration that prepares the body for more intense exercises.

weight: Refers to the plates or weight stacks, or the actual poundage on the bar or dumbbell.

weightlifting belt: A large leather support worn around the waist by bodybuilders to provide support to the lower-back muscles.

wrist straps: Thick bands of various materials worn to support the wrists, allowing weight lifters to lift heavier weights.

GLOSSARY

LATIN TERMS

The following glossary explains the Latin terminology used to describe the body's musculature. Certain words are derived from Greek, which has been indicated in each instance.

CHEST

coracobrachialis: Greek *korakoeidés*, "ravenlike," and *brachium*, "arm"

pectoralis (major and minor): *pectus*, "breast"

ABDOMEN

obliquus externus: *obliquus*, "slanting," and *externus*, "outward"

obliquus internus: *obliquus*, "slanting," and *internus*, "within"

rectus abdominis: *rego*, "straight, upright," and *abdomen*, "belly"

serratus anterior: *serra*, "saw," and *ante*, "before"

transversus abdominis: *transversus*, "athwart," and *abdomen*, "belly"

NECK

scalenus: Greek *skalénós*, "unequal"

semispinalis: *semi*, "half," and *spinae*, "spine"

splenius: Greek *spléníon*, "plaster, patch"

sternocleidomastoideus: Greek *stérnon*, "chest," Greek *kleís*, "key," and Greek *mastoeidés*, "breastlike"

BACK

erector spinae: *erectus*, "straight," and *spina*, "thorn"

latissimus dorsi: *latus*, "wide," and *dorsum*, "back"

multifidus spinae: *multifid*, "to cut into divisions," and *spinae*, "spine"

quadratus lumborum: *quadratus*, "square, rectangular," and *lumbus*, "loin"

rhomboideus: Greek *rhembesthai*, "to spin"

trapezius: Greek *trapezion*, "small table"

SHOULDERS

deltoideus (anterior, medial, and posterior): Greek *deltoeidés*, "delta-shaped"

infraspinatus: *infra*, "under," and *spina*, "thorn"

levator scapulae: *levare*, "to raise," and *scapulae*, "shoulder [blades]"

subscapularis: *sub*, "below," and *scapulae*, "shoulder [blades]"

supraspinatus: *supra*, "above," and *spina*, "thorn"

teres (major and minor): *teres*, "rounded"

UPPER ARM

biceps brachii: *biceps*, "two-headed," and *brachium*, "arm"

brachialis: *brachium*, "arm"

triceps brachii: *triceps*, "three-headed," and *brachium*, "arm"

LOWER ARM

anconeus: Greek *anconad*, "elbow"

brachioradialis: *brachium*, "arm," and *radius*, "spoke"

extensor carpi radialis: *extendere*, "to extend," Greek *karpós*, "wrist," and *radius*, "spoke"

extensor digitorum: *extendere*, "to extend," and *digitus*, "finger, toe"

flexor carpi pollicis longus: *flectere*, "to bend," Greek *karpós*, "wrist," *pollicis*, "thumb," and *longus*, "long"

flexor carpi radialis: *flectere*, "to bend," Greek *karpós*, "wrist," and *radius*, "spoke"